Rosemary Janine Arnold F9

The Institute of Biology's
Studies in Biology no. 13

Understanding the Chemistry of the Cell

by Geoffrey R. Barker Ph.D., D.Sc.
Professor of Biological Chemistry, University of Manchester

Edward Arnold (Publishers) Ltd

First published 1968
Reprinted 1970

Boards edition SBN : 7131 2211 0
Paper edition SBN : 7131 2212 9

Printed in Great Britain by
William Clowes and Sons Ltd, London and Beccles

General Preface to the Series

It is no longer possible for one textbook to cover the whole field of Biology and to remain sufficiently up to date. At the same time students at school, and indeed those in their first year at universities, must be contemporary in their biological outlook and know where the most important developments are taking place.

The Biological Education Committee, set up jointly by the Royal Society and the Institute of Biology, is sponsoring, therefore, the production of a series of booklets dealing with limited biological topics in which recent progress has been most rapid and important.

A feature of the series is that the booklets indicate as clearly as possible the methods that have been employed in elucidating the problems with which they deal. There are suggestions for practical work for the student which should form a sound scientific basis for his understanding.

1968

INSTITUTE OF BIOLOGY
41 Queen's Gate
London, S.W.7.

Preface

This booklet is intended to give a brief outline of the methods used in studying the chemistry of living systems. New techniques provide detailed information concerning the structures of complex materials such as proteins, carbohydrates and the nucleic acids, and the reactions which they undergo. The nature and sequence of chemical changes in the intact organism can be recognized using isotopically labelled compounds and the environment within the living cell can be at least partly reconstructed in tissue extracts. Cellular reactions are now seen not as the consequence of living processes but as the means whereby such processes operate. Above all, the chemical mechanisms of heredity have provided a clear demonstration of the unity of the biological world.

Manchester, 1968

G.R.B.

Contents

Living Cells and their Components 1

The scientist who strives towards an understanding of the chemistry of the cell is confronted by a perplexing variety in the forms of life. The study of biology is, in part, directed towards the detailed description of differences between one organism and another. One of the achievements of modern biochemistry is the demonstration in chemical terms of the essential unity of life which is often obscured by diversity in outward form.

1.1 The diversity of living cells

Bacteria probably represent the simplest living system capable of independent existence. Although simple in form and showing relatively little diversity of structure, they are no less complex than higher organisms in their chemical constituents and show remarkable capacity for chemical synthesis. For various reasons they are most suitable for chemical study. A pure bacterial culture is a collection of virtually identical cells which has the advantage of chemical uniformity. Using only simple nutrients as starting materials, bacteria perform elaborate chemical syntheses within minutes, and within a matter of hours, the number of cells in a culture, and consequently all the chemical constituents of the cells, may increase a thousandfold.

Prodigious though the synthetic capacity of a bacterial cell may be under favourable conditions, it is very much at the mercy of its environment. It depends on the free diffusion of nutrients into the cell and has little defence against the entry of unwanted material. At the other end of the scale, in man and other animals, food absorption is confined to certain cells only and, in effect, only the desirable parts of the environment are made to come into contact with these cells through the act of feeding. Other examples of cell specialization include the absorption of oxygen from the air by blood cells in the lung and the excretion of unwanted products through the kidneys. All these functions are fulfilled by all the cells in a culture of a unicellular organism. It follows that in more highly organized creatures, the chemical processes associated with a particular type of cell will depend on the tissue of which it is a part.

When teased out of a piece of tissue, both plant and animal cells can often be grown in nutrient medium in much the same way as cells of a unicellular organism, and such 'cultured' tissue provides useful homogeneous populations of cells. However, cells grown under such conditions quickly lose their specialized properties and become 'de-differentiated'.

It is clear that the biochemist must be careful in his choice of material for study. In addition to the distinction between unicellular and multi-cellular forms, some types of organism are capable of performing chemical operations which are not represented at all in others. This is particularly true of the photosynthetic processes in green plants. Green plants differ in a more general way from most other types of organism in being autotrophic: that is, they are able to synthesize all their cellular constituents from inorganic materials. By contrast, the large majority of living organisms, termed heterotrophic, require at least part of their food in the form of organic material. Thus heterotrophs are ultimately dependent on autotrophs, which suggests that the first forms of life were autotrophic. This possibility poses many questions concerning biochemical evolution which can be only partially answered at present.

1.2 Organic compounds and their origin

Before it is possible to understand how chemical reactions in the cell serve the processes of life, it is necessary to know in some detail the nature of the materials concerned. Knowledge of the chemical constituents of living cells clearly sets a limit to the understanding of cellular reactions.

Certain types of compound are invariably found in all living cells, which suggests that they are associated with processes vital to all forms of life. In fact, the question can be asked whether the natural existence of these compounds is a direct consequence of living processes. Up to about the middle of the last century scientists believed this to be so, and went further in thinking it impossible for such compounds to be made outside the living cell. However, not only can characteristic components of living organisms be synthesized in the laboratory, but it now appears that at least some could have arisen naturally other than in association with biological processes. Thus it is possible that many materials which we normally regard as of biological origin may have existed on the earth before the advent of life. A number of types of compound such as amino acids, purines and pyrimidines, which are universally found in living organisms, have been shown to be formed under conditions which could well have obtained on the primitive earth. The hypothesis has been put forward, and is gaining ground, that such materials were first produced and reproduced in the non-living world. It is not inconceivable that complex chemical structures similar to those now found associated with living cells arose initially through some form of chemical evolution which antedated biological evolution.

Proteins, Carbohydrates and Nucleic Acids 2

In spite of the diversity of the structure of living cells, the study of their constituents reveals a remarkable degree of uniformity. Of particular note is the universal association with living material of three types of large molecule, namely the proteins, polysaccharides and nucleic acids. While differing in detail from one organism to another, these macromolecules fulfil similar functions in all, and it is of interest to consider their chemical basis.

2.1 Amino acids, peptides and proteins

The amino acids possess both amino and carboxyl groups and exhibit basic and acidic properties:

$$H_2N\cdot CHR\cdot CO_2H$$

In the simplest, namely glycine, R is hydrogen, but over twenty natural amino acids are known in which R may be a hydrocarbon residue, or may carry various functional groups such as $\cdot NH_2$, $\cdot CO_2H$, $\cdot OH$, $\cdot SH$ and others.

The carboxyl group confers acidic properties because it readily loses a proton to give the negatively charged carboxylate ion; conversely, the basic amino group can react with a proton to give a positively charged ion containing a quadricovalent nitrogen atom. Both these processes are reversible and the position of equilibrium will depend on the concentration of protons in the solution. For the present purpose, the hydration of the proton can be ignored. In pure water, the molar concentration of protons is 10^{-7}, usually expressed as pH 7, where pH is defined as $-\log_{10}$ (proton concentration). If the concentration of protons is relatively high (e.g. 10^{-1} M; pH=1) they will tend to react with carboxylate ions to give uncharged $\cdot CO_2H$ groups and with uncharged amino groups to give $\cdot \overset{+}{N}H_3$ ions. If the concentration of protons is low (e.g. 10^{-10} M; pH=10), the opposite will be the case and the groups just referred to will revert to carboxylate ions and uncharged amino groups respectively. At some intermediate value of pH, the tendencies for the $\cdot CO_2H$ group to lose a proton and for the $\cdot NH_2$ group to accept one will be equal and both will become charged:

$$\underset{\text{Low pH}}{H_3\overset{+}{N}\cdot CHR\cdot CO_2H} \rightleftharpoons H_3\overset{+}{N}\cdot CHR\cdot CO_2^- \rightleftharpoons \underset{\text{High pH}}{H_2N\cdot CHR\cdot CO_2^-}$$

Thus, by virtue of the presence of both amino and carboxyl groups in the same molecule, the unionized form of an amino acid is present in very small amount in aqueous solution.

Combination between the carboxyl group of one amino acid molecule and the amino group of another results in the formation of the same type of linkage as in amides:

$$H_2N \cdot CHR \cdot CO \cdot NH \cdot CHR \cdot CO_2H$$

Such compounds are called peptides, and, because of the bifunctional nature of amino acids, can contain large numbers of molecules linked together in this way. Naturally occurring peptides are known with two (dipeptides), three (tripeptides) and more amino acid residues in the chain. The melanocyte-stimulating hormone (MSH), which is concerned with pigment formation in man, is a polypeptide containing twenty-two amino acid residues. Indeed the MSH found in other mammals has a very similar sequence of amino acid residues in the polypeptide to that found in man. This implies a close relationship between structure and physiological function, and also between structure and evolutionary development.

The proteins are much larger molecules built up from polypeptides. The molecule of bovine insulin contains fifty-one amino acid residues and that of human haemoglobin many hundreds; in others, such as the proteins of seeds, about which information is, as yet, far from complete, the number of residues is much larger. In the first place the proteins will be considered as large polypeptides.

From the general structure of a peptide, it is clear that one amino group and one carboxyl group is free at the ends of the chain. Some peptides are known, such as the antibiotics gramicidin S and tyrocidin, in which a cyclic structure is present, formed by interaction between terminal groups. In the majority of cases, however, these groups are free and behave in the same way as regards basic and acidic properties as their counterparts in a simple amino acid. In addition, the R groups of some amino acids carry amino or carboxyl groups as in lysine and glutamic acid respectively:

$$\begin{array}{l} H_2N \cdot CH \cdot CO_2H \\ \quad\quad\;\; \dot{C}H_2 \cdot CH_2 \cdot CH_2 \cdot CH_2 \cdot NH_2 \end{array} \qquad \begin{array}{l} H_2N \cdot CH \cdot CO_2H \\ \quad\quad\;\; \dot{C}H_2 \cdot CH_2 \cdot CO_2H \end{array}$$

Lysine — Glutamic acid

The presence of basic and acidic groups in protein molecules is most important in connection with one of their functions in the cell. This can best be understood by considering a simple amino acid molecule in solution.

In an aqueous solution of hydrochloric acid the concentration of protons is high and the pH consequently low. If an amino acid is added to such a solution, some of the protons will combine with ionized carboxyl groups of the amino acid to give the unionized form, and with amino groups to give the corresponding ion. Both these processes reduce the concentration of protons thereby raising the pH. This phenomenon, which is called buffering, plays an important rôle in cellular processes since chemical reactions in living cells take place only within narrow limits of pH. The pH in the proximity of a large protein molecule will be determined partly by the number of basic and acidic groups in its structure and, because of buffering, the pH will normally remain constant within certain limits. Thus it is possible for chemical reactions involving proteins to take place under relatively stabilized conditions.

In addition to the presence of basic and acidic groups in the side-chains, some R groups of natural amino acids carry other functional groups such as hydroxyl, in, for example, serine and tyrosine, or thiol in cysteine:

$$H_2N \cdot \underset{\cdot}{C}H \cdot CO_2H$$
$$\dot{C}H_2 \cdot OH$$

Serine

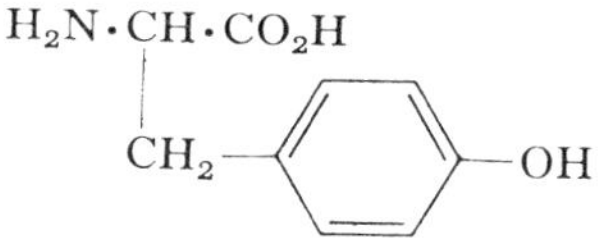

$$H_2N \cdot CH \cdot CO_2H$$
$$|$$
$$CH_2 — C_6H_4 — OH$$

Tyrosine

$$H_2N \cdot \underset{\cdot}{C}H \cdot CO_2H$$
$$\dot{C}H_2 \cdot SH$$

Cysteine

The thiol groups of cysteine residues play a very special rôle in many proteins for thiols are readily oxidized to disulphides in which two sulphur atoms are joined together and it is not surprising that such linkages are formed between cysteine residues in polypeptides:

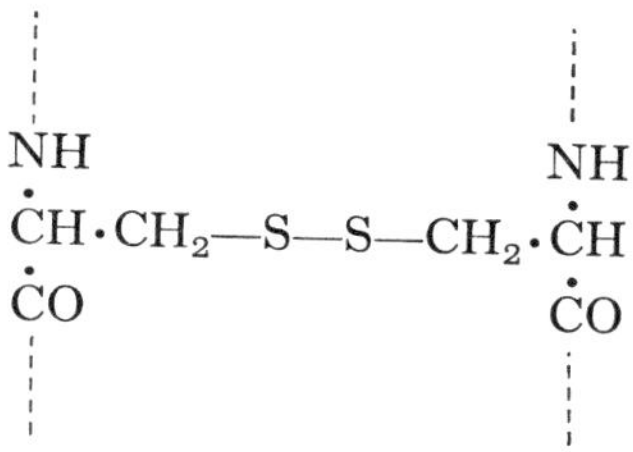

$$
\begin{array}{lcr}
\vdots & & \vdots \\
NH & & NH \\
\dot{C}H \cdot CH_2—S—S—CH_2 \cdot & & \dot{C}H \\
\dot{C}O & & \dot{C}O \\
\vdots & & \vdots
\end{array}
$$

The formation of this type of structure is most important in connection with the gross structure of large protein molecules. In the first place, it may result in the joining together of adjacent polypeptide chains to form bundles. Secondly, protein molecules are so large that the long polypeptide chains may bend back on themselves in such a way that a disulphide bridge can be formed between two cysteine residues in the same chain. Many of the interactions of proteins with other molecules which take place during cellular processes depend on the existence of a very specific shape or conformation of the protein molecule. Disulphide bridges play a central rôle in producing and preserving such conformations.

Size alone does not confer on proteins their special properties. Shape is also of prime importance. Moreover, the covalent structure, or primary structure as it is called, of a protein, including the covalent disulphide linkages is not alone in determining the more subtle features of a protein in relation to its biological function. A major factor controlling the overall shape of a protein molecule is the presence of a large number of linkages much weaker in character than covalent bonds. These are the so-called hydrogen bonds formed by the sharing of hydrogen atoms in the following way:

```
        \             /
         C=O--H—N
        /          \
R·CH                  CH·R
        \          /
         N—H--O=C
        /             \
```

In order that as many bonds of this type as possible can be formed within a single polypeptide chain, the chain must be folded in a particular way, which has been termed the α-helix and is shown in Fig. 2–1. Alternatively,

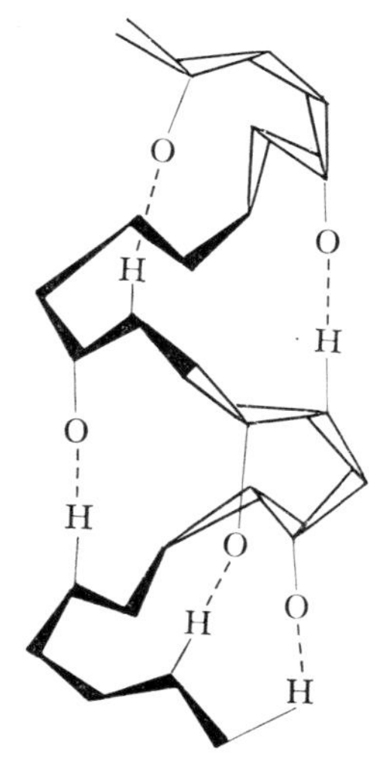

Fig. 2–1 The α-helix structure of a protein. The main bonds of the polypeptide chain are shown together with hydrogen bonds between carbonyl oxygen and amide hydrogen atoms. Shaded loops of the coil protrude forwards; open loops project backwards.

hydrogen bonds may bind adjacent polypeptide chains together to form what is called a pleated sheet structure, but which is better described by likening it to a venetian blind. These structures clearly involve a high degree of orderliness and regularity and are referred to as secondary structures. Such detail is only revealed by measuring the positions of all the atoms in the polypeptide chain relative to each other using the diffraction of x-rays.

Orderliness in the structures of proteins has its origin in the amino acids represented in the polypeptide chains. All except glycine have asymmetric molecules due to the presence of four different groups round the central carbon atom. It is found that all the amino acids present in proteins have the same arrangement of groups, differing only in the nature but not the orientation of the R group. Thus in ordered secondary structures such as the α-helix, the R groups are distributed in space in a regular fashion. It is of interest that in relatively small polypeptides which do not possess secondary structure and in which regularity of structure is consequently of less significance, the same uniformity of configuration among the amino acids is not encountered.

In addition to the ordered coiling or alignment of polypeptide chains in proteins resulting from interactions between —NH— and —CO— residues, further folding, called tertiary structure, occurs and thus accommodates interactions between R groups. Finally, as mentioned above, protein molecules may consist of bundles of polypeptide chains. In insulin, the aggregation of chains may be a consequence of the primary structure, the bundles being tied together by covalent disulphide bridges. In others, such as haemoglobin, the bundles are more loosely bound such that they may be taken apart and reconstituted. The association of sub-units, which may or may not be alike, is referred to as quaternary structure.

During the last fifteen years, the primary structures of proteins have been studied by a variety of methods. The first success came with the establishment of the structure of bovine insulin. The molecule consists of two polypeptides, containing twenty-one and thirty amino acid residues respectively, held together by two disulphide bridges and containing one inter-fold bridge within the smaller peptide. Other structures such as those of haemoglobin and the enzymes ribonuclease and lysozyme have been elucidated and it is likely that the list will grow with increasing speed. The application of computers to the interpretation of x-ray diffraction patterns is already enabling secondary, tertiary and quaternary structures to be established with greater rapidity.

Proteins play leading parts in the chemistry of the cell. Most, if not all of the structural elements of the cell are composed at least in part of protein, which may therefore be said to provide the environment in which the chemical processes of life take place. Above all, in the form of enzymes, they themselves occupy a controlling position in cellular processes by acting as specific catalysts of biological reactions.

2.2 Carbohydrates

The name carbohydrate expresses the fact that typical members of the group, of which glucose ($C_6H_{12}O_6$) is the commonest, have the general formula $C_x(H_2O)_y$. Carbohydrates are as numerous and widespread in occurrence as the amino acids and proteins and fulfil many diverse functions. As with the amino acids and proteins, the carbohydrates are not only present in the structural elements of the cell but also contribute to the life of the cell by virtue of the chemical changes which they undergo. They do not share with proteins the catalytic properties associated with enzymes.

The carbohydrates comprise both small molecules and large: glucose typifies the single unit structures, or monosaccharides, which when linked together with the elimination of the elements of water, give rise to the di-, tri-, etc. and polysaccharides. In contrast to the proteins, the majority of which are built up from numerous different amino acids, polysaccharides for the most part consist of only one, or at the most, two or three, different kinds of monosaccharide joined together.

As can be seen from structure 2-I, glucose consists of a chain of six carbon atoms carrying five hydroxyl groups and one carbonyl function, in this case the aldehyde residue. The name 'aldose' is given to carbohydrates of this type. Other monosaccharides have chains consisting of three, four, five, or even seven carbon atoms, but those which constitute the units of larger molecules invariably have five- or six-carbon chains.

CHO	$CH_2\cdot OH$	CHO	CHO
$CH\cdot OH$	CO	$HO\cdot CH$	$CH\cdot OH$
$HO\cdot CH$	$HO\cdot CH$	$HO\cdot CH$	$HO\cdot CH$
$CH\cdot OH$	$CH\cdot OH$	$CH\cdot OH$	$HO\cdot CH$
$CH\cdot OH$	$CH\cdot OH$	$CH\cdot OH$	$CH\cdot OH$
$CH_2\cdot OH$	$CH_2\cdot OH$	$CH_2\cdot OH$	$CH_2\cdot OH$
2-I	2-II	2-III	2-IV

In the carbohydrate fructose, or fruit sugar (2-II) the aldehyde form of the carbonyl function has been exchanged for the ketone form, a fact reflected in the generic name 'ketose'. A few more variants will be referred to later. Individual monosaccharides differ less frequently in the nature of the functional groups they carry than in their *distribution in space*. Thus mannose (2-III) and galactose (2-IV) differ from glucose only in the distribution of hydroxyl groups round the asymmetric atoms of the carbon chain. This situation is in direct contrast to that found in the amino acids. Whereas the shape of a protein molecule is determined partly by the uniformity in configuration of structurally different amino acids, regularity within the molecules of polysaccharides stems from the repetition of identical monosaccharide units. In order to appreciate fully how monosaccharide units are joined to form polysaccharides it is necessary to examine the nature of the units in more detail.

Although certain of the reactions of glucose are those expected of a molecule having structure 2-I, others are not adequately explained on this basis. The molecule of glucose is known to adopt a number of other structures two of which are represented in 2-V and 2-VI (X=OH). At first sight it is difficult to see that these are different forms of the same

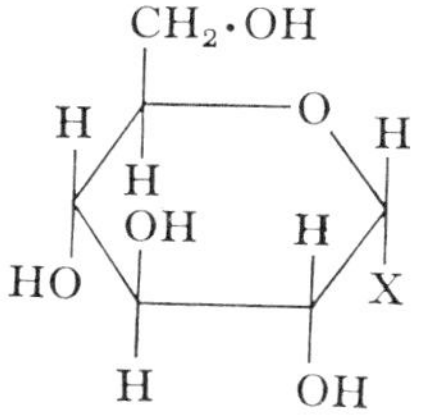

2-V 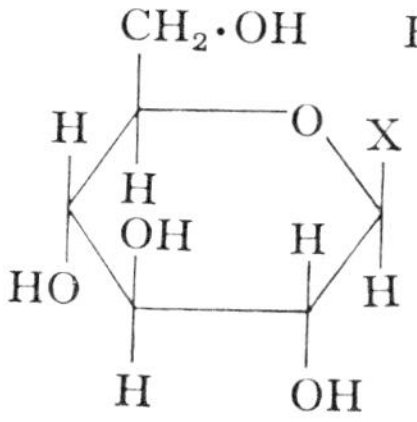 2-VI 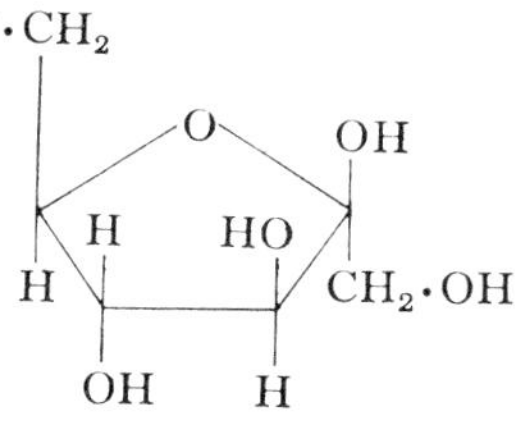 2-VII

molecule. The explanation is to be found in the general reaction of the carbonyl group with hydroxylic compounds illustrated by the formation of acetaldehyde dimethyl acetal:

$$CH_3\cdot CHO + 2CH_3\cdot OH \rightarrow CH_3\cdot CH(OCH_3)_2 + H_2O$$

Structure 2-I possesses both carbonyl and hydroxyl groups in the same molecule, and this allows acetal formation to take place within the molecule. However, since the carbon chain of 2-I cannot be bent in two directions at once, the aldehyde group can react with only one hydroxyl group and acetal formation can proceed only to the half-way stage. Structures 2-V and 2-VI are therefore called hemiacetals. The formation of these structures introduces an extra asymmetric centre into the molecule and consequently two forms exist designated α and β, represented respectively by 2-V and 2-VI. The pattern of diffraction of x-rays by a crystal of ordinary glucose indicates that it possesses structure 2-V. In aqueous solution, a dynamic equilibrium is set up involving all three structures 2-I, 2-V and 2-VI and this is the cause of the spontaneous change of optical rotation which takes place after dissolving crystalline glucose in water.

Structure 2-VII illustrates a hemiacetal form of fructose in which the ketone group has reacted with the hydroxyl group at carbon five producing a five-membered instead of a six-membered ring.

Glucose reacts with methanol to give α- and β-forms of the compound methyl glucoside (2-V and 2-VI; X=OCH_3), in which acetal formation is complete. Compounds of this type are given the generic name 'glycoside' and are the key to understanding the large carbohydrate molecules elaborated by living cells. By virtue of the fact that monosaccharides possess hydroxyl groups, glycosides can be formed between two such molecules as illustrated by sucrose and lactose (milk sugar):

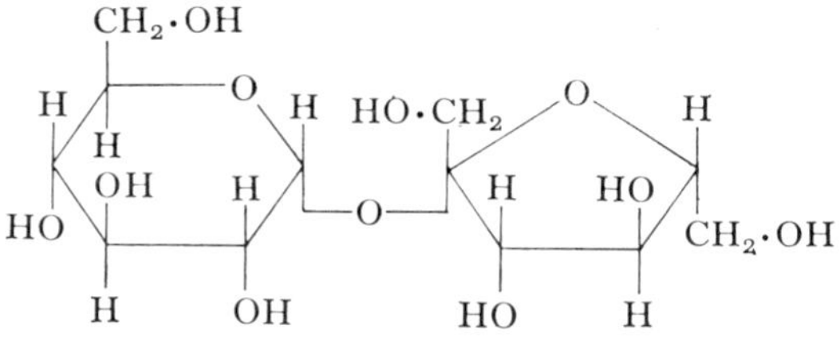

Sucrose

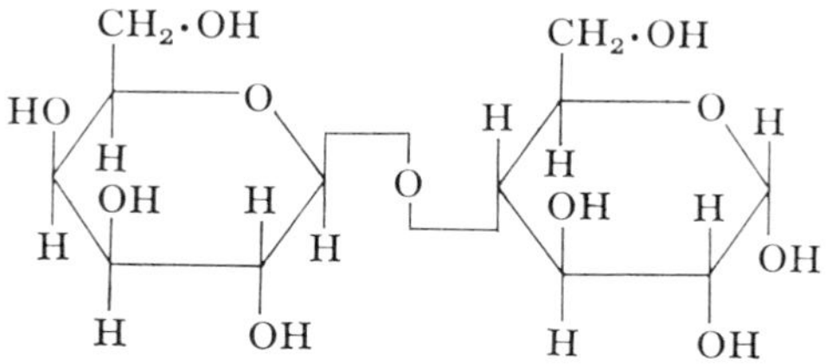

Lactose

In the former case, both the glucose and fructose units achieve the glycoside structure and the potential carbonyl groups of both halves of the molecule are protected. For this reason, sucrose serves as a store of carbohydrate for plant tissues in which it occurs, being less liable to rapid oxidation than the free monosaccharides. In the molecule of lactose, only the left-hand galactose unit is protected in this way, the right-hand glucose unit retaining the hemiacetal structure. This mode of linkage, unlike that in sucrose, is capable of virtually indefinite extension as in the molecule of cellulose which consists of a very large number of glucose units joined together:

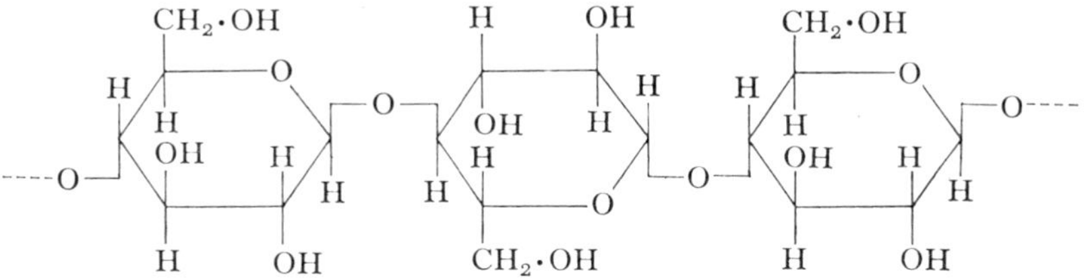

Cellulose is very widespread in skeletal materials of plants, in which high chemical reactivity is not required. Polysaccharides built on the same general plan as cellulose also serve as storage materials in both plant and animal tissues. Starch and glycogen, present respectively in plants and liver, both have structures in which very many molecules of glucose are joined together in chains, which, unlike those in cellulose, are branched:

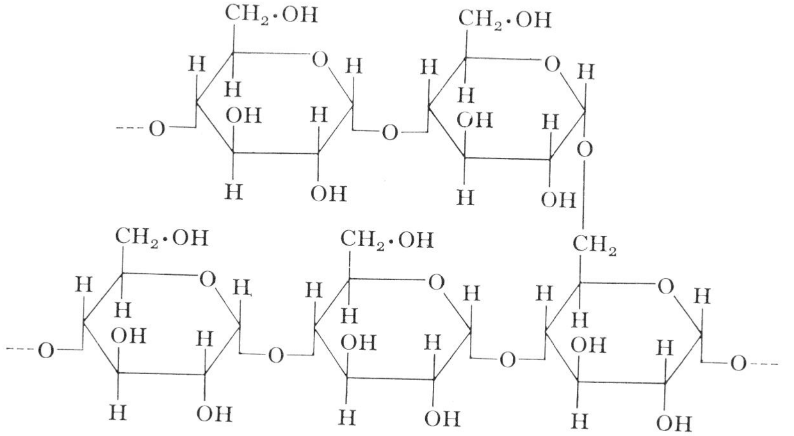

The stored material is made available by unhitching units of glucose from the ends of the chains.

Although glucose, fructose, galactose and a few other monosaccharides are the most commonly found in polysaccharides of plant and animal origin, others fulfil special functions. Thus gel-forming polysaccharides are formed by multiple linking of molecules of uronic acids, such as glucuronic acid (2-VIII) in which the terminal $\cdot CH_2OH$ has been converted into $\cdot CO_2H$. Chitin, a major component of the exoskeletons of arthropods is built similarly from a derivative of glucosamine (2-IX) in which a hydroxyl group is replaced by the amino group. Monosaccharides themselves are also found in various combined forms including simple glycosides of hydroxylic compounds, and, particularly, phosphoric esters such as glucose 6-phosphate (2-X):

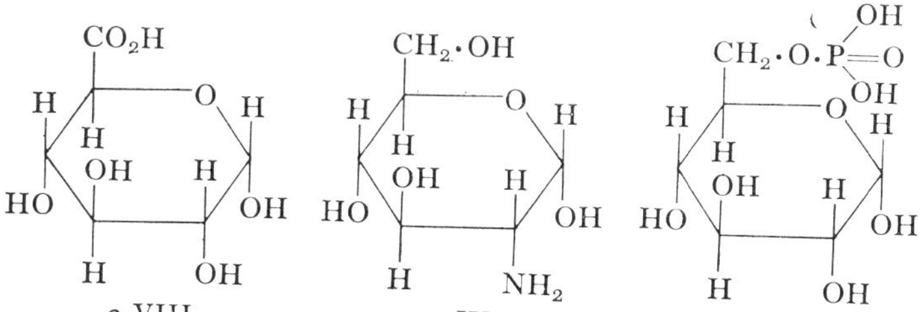

2-VIII 2-IX 2-X

A large number of different materials are elaborated from a few types of monosaccharide unit. This economy in chemical design means that the synthetic mechanisms must operate with faithful reproducibility in order to provide materials playing highly specialized rôles.

2.3 Nucleotides

Nucleotides form the repeating structural unit of the nucleic acids, the third main group of natural macromolecules. The molecule of a nucleotide comprises three chemical subunits, namely, a nitrogen-containing ring compound linked to a carbohydrate residue which in turn is combined with phosphoric acid. Two main families of nucleotides are known, characterized by the presence of the carbohydrates D-ribose and D-2-deoxyribose respectively. Typical examples are adenosine monophosphate (AMP) (2-XI; R=OH) and deoxyadenosine monophosphate (d-AMP) (2-XI; R=H).

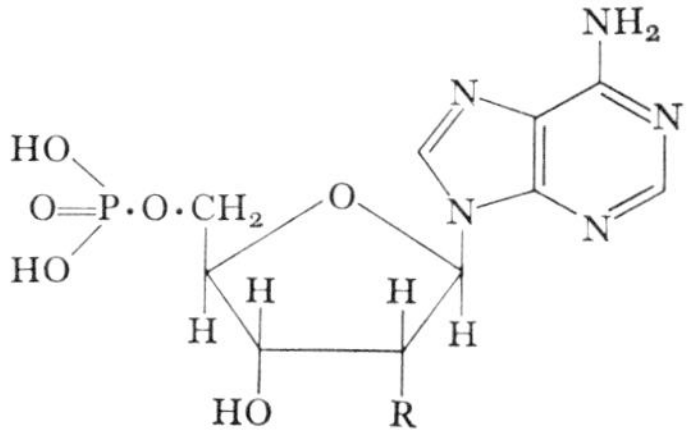

2-XI

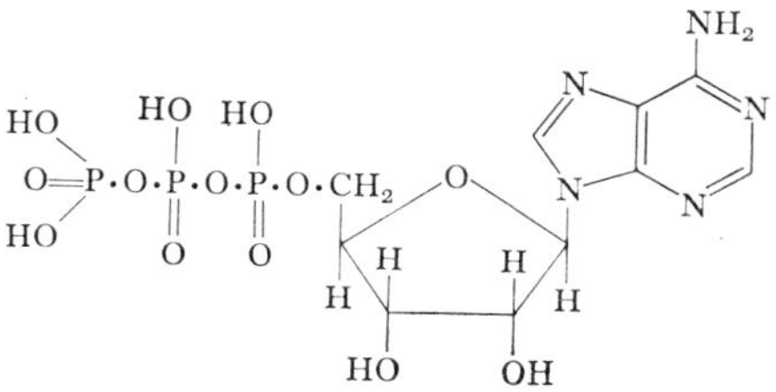

2-XII

Others carry different nitrogenous residues. Besides the simple molecular pattern exemplified by structure 2-XI, nucleotides are present in tissues in which one or two extra phosphoric acid residues are attached as illustrated by the structure 2-XII of adenosine triphosphate (ATP). Nucleotides of this sort undergo a large number of reactions in which one or both of the two extra phosphoric acid groups are lost. In a number of reactions ATP acts as a phosphorylating agent as, for instance, in the conversion of glucose into glucose 6-phosphate. This is achieved by transfer of a phosphate residue from ATP with the simultaneous formation of adenosine diphosphate (ADP). The energy liberated by hydrolytic removal of the phosphate residue from ATP is used in forming the new bond with glucose. As will be seen later, it is the free energy available when ATP or ADP undergoes hydrolysis that enables these compounds to perform so many useful functions in the cell.

Nucleotides carrying three phosphoric acid residues also serve as precursors of polynucleotides, pyrophosphate being split off during the polymerization. Figure 2–2 illustrates the way in which nucleotides are joined together in the form of polynucleotides or nucleic acids. Polynucleotide chains are always built up from either ribonucleotides or deoxyribonucleotides but never contain both types in one chain. Thus it is possible to distinguish clearly two distinct types of nucleic acid, namely ribonucleic acid (RNA) (R=OH) and deoxyribonucleic acid (DNA) (R=H). It will be seen later that these have quite distinct biological functions. DNA and RNA differ not only in the nature of the carbohydrate residues but also in the bases which characterize the constituent nucleotides. Apart from small proportions of other bases, RNA contains nucleotide units derived from adenine, guanine, cytosine and uracil. DNA contains the deoxyribonucleotides derived from adenine, guanine and cytosine, but thymine replaces uracil in the fourth nucleotide. The structure depicted is only a small segment of a very long chain. Three different types of RNA can be distinguished, the two smaller molecules comprising approximately 80 and 300 nucleotide units respectively; the largest RNA molecules contain up to about 3500 residues. It is difficult to define the size of DNA molecules except to say that they are very big indeed. Claims have been made for molecules of DNA as big as 10^6 nucleotide units, but the molecular weight of DNA is dependent to a considerable degree on the method of preparation. Such a long thin molecule is easily mechanically torn apart into shorter lengths by agitation or stirring. Although it is impossible to assess with any degree of certainty the molecular size of DNA as it is in the cell, it is very large and outstrips in size all other cell constituents by a large margin.

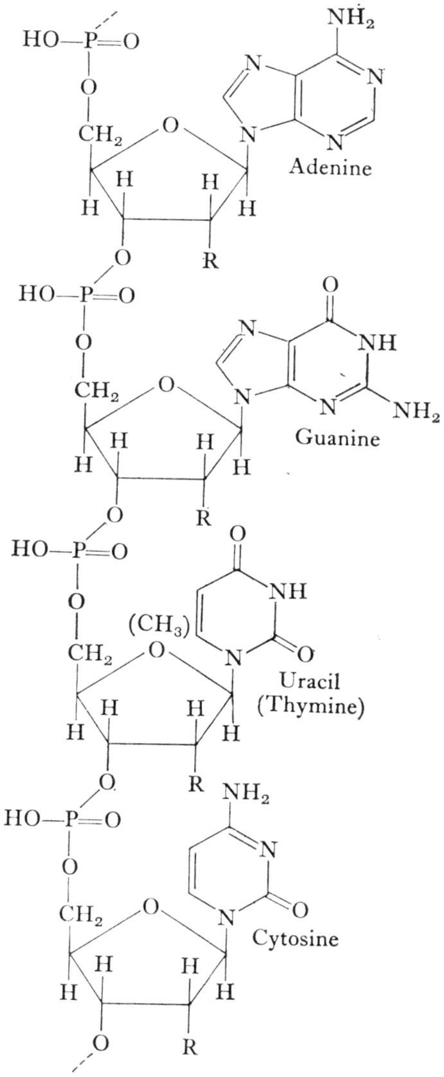

Fig. 2–2 The type of structure present in nucleic acids. RNA contains uracil and has R=OH; DNA contains thymine and has R=H.

As with the proteins, native nucleic acid molecules are held together by weak hydrogen bonds as well as the covalent bonds which define the primary structure. Hydrogen bonds in the nucleic acids are formed between pairs of basic residues, cytosine being linked to guanine and adenine to either uracil or thymine. It can be seen from Fig. 2–3 that such pairing involves a very specific spatial arrangement of the groups. The first hint

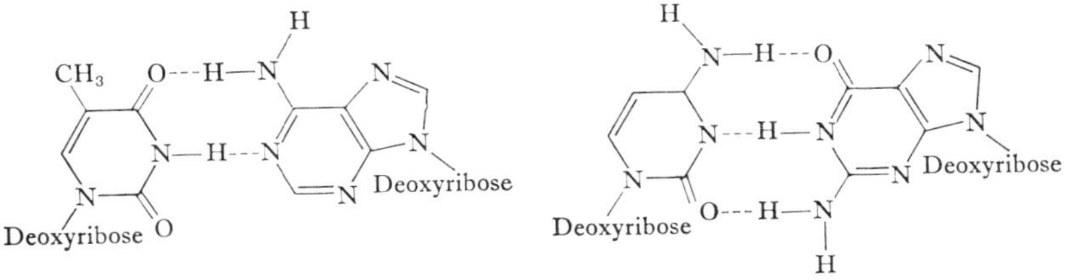

Fig. 2–3 The method of hydrogen bonding between pairs of bases in DNA.

that pairing occurs between bases in this way was that analyses of DNA isolated from a wide variety of sources revealed the universal presence of adenine and thymine, and guanine and cytosine in equimolecular amounts. This is expressed most simply in the form:

$$A = T; \qquad G = C$$

where A, T, G and C represent the molecular proportions of the four bases. Variations in composition between samples of DNA of different origins are confined to differences in $A+T$ relative to $G+C$. It was then found that the diffraction of x-rays by DNA could be most readily interpreted by assuming a double helical structure in which adenine and thymine residues and guanine and cytosine residues are adjacent to each other in the two spirals illustrated in Fig. 2–4. Clearly, the existence of cross-

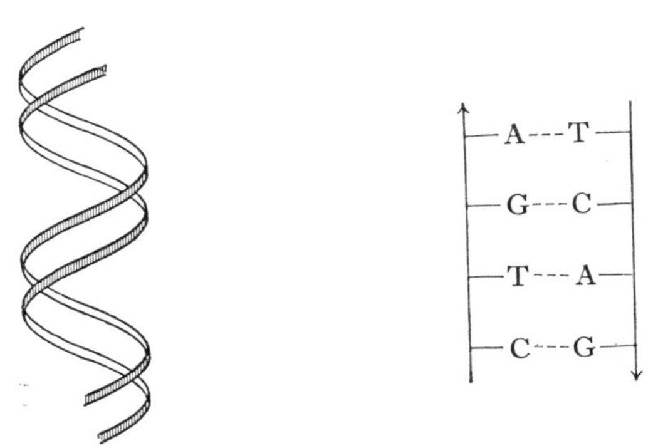

Fig. 2–4 Double helix illustrating the axes of two polynucleotide chains in the DNA molecule.

Fig. 2–5 The pairing of bases by hydrogen bonding between complementary strands of DNA in the double helix.

linkages of this type throughout the double helical structure depends on the complementary nature of the two strands and this is seen most easily by ignoring for the moment the spiral structure and visualizing the sequence of nucleotide residues in the two strands of the molecule. In Fig. 2–5 the left-hand strand is the same as that in Fig. 2–2, the direction of the arrowhead indicating the sense of the deoxyribosephosphate backbone. The backbone of the right-hand half is orientated in the opposite sense, and the sequence of nucleotides is seen to be complementary to that of the other strand. Almost all native DNA is found to have this type of double secondary structure and, although the precise mechanism of the process is not clear, it is believed that the synthesis of DNA in the cell involves the alignment of nucleotides on the temporarily separated strands prior to their polymerization as illustrated in Fig. 2–6. In this way, the DNA molecule is seen to be self-replicating, a property which would be difficult to visualize in a less ordered structure.

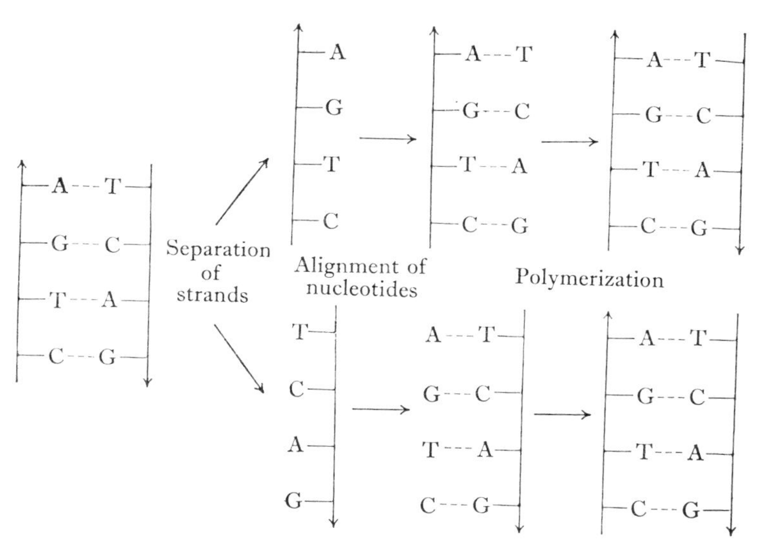

Fig. 2–6 Diagrammatic representation of the replication of DNA.

Extracts of cells can be made which will synthesize RNA from a mixture of nucleoside triphosphates, provided that DNA is present. Moreover, the composition of the RNA which is formed is similar to that of DNA, which suggests the existence of a process analogous to that shown in Fig. 2–6 but differing in that the nucleotides aligned are ribonucleotides derived from adenine, guanine, cytosine and uracil, adenine pairing with uracil instead of thymine.

The pairing of bases in adjacent nucleotide chains is most important in the synthesis of cell components. The mechanism whereby the DNA molecule is replicated ensures faithful reproduction in every detail. This process together with the formation of a chain of RNA on a template of DNA form the basis of the chemistry of heredity which is discussed later.

Methods of Studying Chemical Reactions of the Cell 3

3.1 Measurement of respiration

It has already been pointed out that the formation of a new bond between two atoms is accompanied by the uptake of energy. Reactions which lead to the linking of atoms together are therefore called endergonic. The necessary energy is normally supplied by the simultaneous breaking of some other bond, that is by linking the endergonic reaction to one which releases energy, namely an exergonic reaction. The transfer of phosphate from ATP to glucose discussed on p. 12 illustrates this. Clearly, the creation of large molecules such as proteins, polysaccharides and polynucleotides involves many endergonic reactions and it is essential that all cells have access to a continuous supply of energy for chemical synthesis. In most organisms this energy is derived from the oxidation of carbon compounds present in foodstuffs with the production of carbon dioxide and the consumption of molecular oxygen. This process is known as respiration and, although associated with the whole organism, is essentially a cellular activity.

One of the most frequent food molecules to serve as an energy source is glucose, the overall oxidation of which is represented as follows:

$$C_6H_{12}O_6 + 6O_2 \rightarrow 6CO_2 + 6H_2O + \text{Energy}$$

The ratio of the number of moles of carbon dioxide produced to the number of moles of oxygen consumed is called the respiratory quotient (R.Q.):

$$\text{R.Q.} = \frac{\text{Volume of } CO_2 \text{ produced at S.T.P.}}{\text{Volume of } O_2 \text{ consumed at S.T.P.}}$$

It can be seen from the equation that for the oxidation of glucose, R.Q. = 1. If some other carbon compound such as glycerol is provided as food, it can be seen from the following equation that R.Q. = 0·857:

$$2C_3H_8O_3 + 7O_2 \rightarrow 6CO_2 + 8H_2O + \text{Energy}$$

A simple calculation shows that for the complete oxidation of a long chain fatty acid such as stearic acid ($C_{18}H_{36}O_2$) R.Q. = 0·69. Thus whereas measurements of the rate of production of carbon dioxide or the rate of uptake of oxygen gives an indication of the extent of utilization of food by an organism, the ratio of these quantities (i.e. R.Q.) helps to identify the nature of the foodstuff undergoing oxidation.

The consumption of oxygen and/or the production of carbon dioxide can be measured manometrically in an apparatus, developed by Otto Warburg, shown in Fig. 3–1. Measurement of the rate of consumption of oxygen by a culture of bacteria illustrates the way in which it is used.

A suspension of the bacteria is placed in the main compartment of the flask, and a solution of the foodstuff, e.g. glucose, is placed in the side-arm. The centre compartment contains sodium hydroxide solution supported on a wick of filter paper to absorb carbon dioxide. Oxygen or air may be introduced through T_1 and taps T_1 and T_2 are then closed. The flask is first brought to equilibrium in a thermostat and then the solution is tipped from the side-arm into the main compartment. The volume is maintained constant by adjusting the screw-clip and the consumption of oxygen is calculated from the decrease in pressure. The rate of consumption of oxygen measured in this way is expressed as Q_{O_2}, the number of microlitres of oxygen at standard temperature and pressure taken up per hour per milligram dry weight of tissue. A similar experiment in which sodium hydroxide is omitted from the centre well, will give measurements from which the rate of production of carbon dioxide can be deduced, and the R.Q. calculated. Full details of the use of manometric apparatus are described by CLARK (see Further Reading).

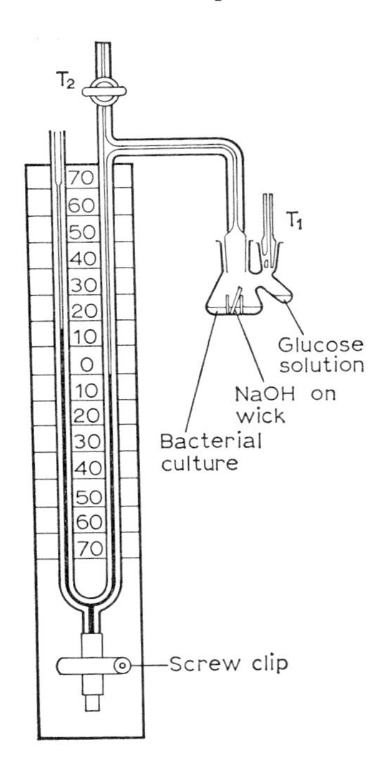

Fig. 3–1 Diagrammatic representation of Warburg apparatus.

This apparatus is most appropriate for the study of suspensions of micro-organisms and is clearly unsuitable for examining whole organisms of any size. However, with multi-cellular organisms, the measurement of Q_{O_2} and R.Q. for the whole organism does not give very useful information about the chemical reactions taking place, because the result is the average of very many different values. Different tissues in a complex organism are likely to have access to slightly different mixtures of food materials at any given time and are likely to need to use what food they have at very different rates. For these reasons, more meaningful results in terms of cell chemistry are obtained by using slices of fresh tissue suspended in a nutrient solution.

3.2 The use of isotopic tracers

The manometric measurement of respiration provides a method of studying the overall oxidation of a compound such as glucose without revealing anything concerning intermediate stages in the process. Isotopic tracers are among the most powerful tools for investigating what is known as intermediary metabolism.

With a few exceptions such as sodium, phosphorus and iodine, the principal elements found in living cells contain a small proportion of isotopic atoms which differ from normal in atomic weight. Table 1 shows some examples.

Table 1 The natural abundance of some stable isotopes.

Element	*Mass number*		*Atom % of isotope in natural element*
	Normal	*Isotope*	
Hydrogen	1	2	0·01
Carbon	12	13	1·1
Nitrogen	14	15	0·37
Oxygen	16	18	0·20

These isotopic atoms have stable nuclei and the proportion of the isotope present remains constant. Moreover, since the proportion of each isotope is the same in materials of biological origin as in non-living materials, it follows that living cells do not distinguish between the different atoms. Gaseous molecules containing isotopic heavy atoms diffuse more slowly than normal molecules and, making use of this fact, it is possible to separate the molecules containing isotopic atoms. These molecules are then recognizable by the increased percentage of isotopic atoms which can be measured using a mass spectrometer. Thus the fate of 'isotopically labelled' compounds which have been synthesized from materials enriched in a given isotope can be traced after being administered to an animal or other organism. By this method, it was shown, for instance, that glycine labelled with the heavy isotope of nitrogen, ^{15}N, after being injected into pigeons, gave rise to uric acid in the excreta having more than the normal percentage of ^{15}N. Furthermore, administration of glycine labelled with ^{13}C also resulted in the excretion of labelled uric acid. Thus it follows that glycine contributes both nitrogen and carbon atoms in the formation of uric acid by the pigeon.

The natural stable isotope of nitrogen is frequently used in elucidating the metabolic interconversion of nitrogenous materials. However, the discovery of artificially produced radioactive isotopes has considerably extended the possibilities of isotopic tracer techniques and simplified the methods of detection. These isotopes have unstable nuclei which disintegrate exponentially with time and emit charged particles or ionizing radiations which are readily measured. The use of radioactive isotopes in biochemical studies is not without its difficulties: care has to be taken that the radiation emitted does not damage cell components, and, with some isotopes, the radioactive decay is so rapid that only short-term experiments are possible. In spite of these drawbacks, radioactive techniques have enabled problems to be solved which could not have been tackled by any other method. Table 2 lists the major isotopes used.

Table 2 The properties of some radioactive isotopes used in biochemistry.

Element	*Mass number*	*Half-life*
Hydrogen (Tritium)	3	12·5 yr
Carbon	14	5760 yr
Sodium	24	15 hr
Phosphorus	32	14·3 day
Sulphur	35	87·1 day
Potassium	42	12·5 hr
Calcium	45	160 day
Iron	59	45 day
Iodine	131	8 day

In a series of consecutive reactions such as:

$$A \rightarrow B \rightarrow C \rightarrow D ------ \rightarrow Z$$

a steady state is reached such that the concentration of each intermediate remains constant. Furthermore, if an isotopically labelled sample of A is introduced into the system, in time, the degree of labelling of each compound in the chain will become the same. However, if labelled A is added and the whole reaction sequence is stopped by suitable means, then it would be found that the concentration of isotope would be highest in A, slightly less in B, still less in C and so on down the line. Thus by means of what is referred to as a 'pulse labelling' experiment, it is possible to discover not only what compounds participate in a chain of reactions, but also the sequence in which they are formed.

If a labelled sample of B is added as a single shot, the isotopic content of B in the mixture will be suddenly increased. As B is converted into C and A into B, this high level of isotope in B will diminish, and that of C will increase. After a time, the isotope will begin to disappear from C, as it is passed on to D. If the isotopic contents of B and C measured at varying intervals of time are plotted against time, the graphs are of the type shown in Fig. 3–2. It is to be noticed that the descending part of the graph

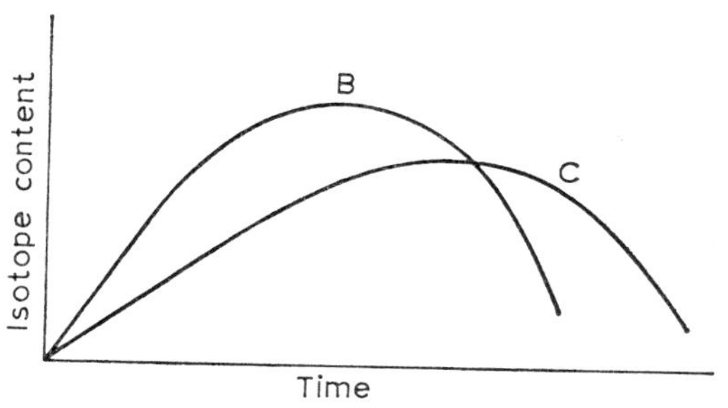

Fig. 3–2 The time course of the incorporation of isotope into precursor and product in a metabolic sequence of reactions.

referring to B passes through the maximum of the curve for C. It has been shown mathematically that this occurs only if B is related to C as precursor to product. In this way it is possible to establish sequences of reactions which would not otherwise have come to light.

3·3 Inherited mistakes in metabolism

The chemical reactions of the cell are so geared together that mostly, under normal circumstances, the product of one reaction serves as the starting material for another. Thus it is difficult to do more than recognize the disappearance of nutrients and the appearance of the finished product of a series of consecutive reactions, whether this is in the form of new tissue or waste products. By such experimental tricks as pulse labelling, it is possible to recognize some intermediate stages in a sequence of reactions, but this is occasionally complicated by the existence of numerous sequences operating simultaneously. In certain inherited diseases, there is a failure of the body to produce an enzyme which normally catalyses a particular reaction in a metabolic sequence. Under these conditions, the product of the reaction preceding the block accumulates and is usually excreted because of the lack of mechanism for its further metabolism. In this way, it is possible to recognize metabolic intermediates which otherwise might escape detection.

During the last century, a rare disease known as alkaptonuria was recognized by virtue of the darkening of the patients' urine on exposure to air. This was later shown to be due to the presence in the urine of the readily oxidizable compound homogentisic acid (see structure on page 22). The condition was found to be aggravated by feeding tyrosine and it became clear that homogentisic acid is an intermediate in the normal breakdown of dietary tyrosine. This fact, which has been confirmed by isotopic experiments would certainly not have been anticipated from a comparison of the structures of tyrosine and homogentisic acid.

Another rare disease, known as pentosuria, is accompanied by the

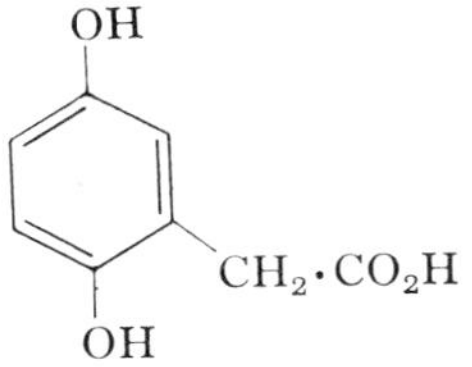

OH (benzene ring with OH, $CH_2{\cdot}CO_2H$, OH)

Homogentisic acid

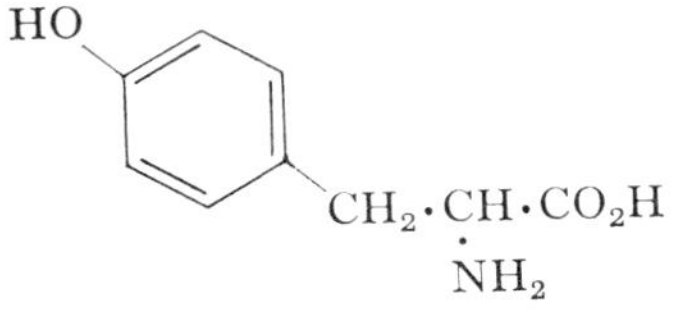

HO (benzene ring with $CH_2{\cdot}CH{\cdot}CO_2H$, NH_2)

Tyrosine

excretion of the pentose sugar L-xylulose. A most interesting fact, which emphasizes the heritable nature of the disease is that it is confined largely to people of Jewish descent. By the same reasoning as above, it appeared that L-xylulose is an intermediate which is normally metabolized further. It was already known that the phosphoric ester of D-xylulose is involved in the utilization of D-glucose, but the excretion of L-xylulose by pentosurics suggested the existence of another pathway. As a result, it was discovered that D-glucose is converted into both D-xylulose and L-xylulose and that the stereochemical structure of xylulose can be reversed through the intermediate formation of the compound xylitol which is optically inactive because it has a symmetrical structure:

D-Xylulose	L-Xylulose
$CH_2{\cdot}OH$	$CH_2{\cdot}OH$
CO	CO
$HO{\cdot}CH$	$CH{\cdot}OH$
$CH{\cdot}OH$	$HO{\cdot}CH$
$CH_2{\cdot}OH$	$CH_2{\cdot}OH$

$CH_2{\cdot}OH$	$CH_2{\cdot}OH$
$CH{\cdot}OH$	$HO{\cdot}CH$
$HO{\cdot}CH$	$CH{\cdot}OH$
$CH{\cdot}OH$	$HO{\cdot}CH$
$CH_2{\cdot}OH$	$CH_2{\cdot}OH$

Two molecules of xylitol, seen to be identical if turned in the plane of the paper.

These examples show that the study of genetic diseases can throw light on the nature of chemical reactions obtaining in the body. Knowledge of such reactions may also point to methods for the alleviation of the conditions accompanying the disease. A serious difficulty is often the recognition of the existence of the metabolic error before it proves lethal but the use of biochemical analysis in diagnosis shows a steadily increasing reward.

For purposes of studying cellular reactions, genetic abnormalities in bacteria and other micro-organisms are the most useful. Many mutant bacteria can be separated from natural populations and, in addition, mutation can be induced by irradiation with ultra-violet light or by chemical agents. Most bacteria are able to synthesize all the nitrogen-containing materials they require, including amino acids and nucleotides, if they are supplied with ammonium salts and an organic source of carbon such as glucose. Bacteria which have an inherited lack of an enzyme involved in one stage of a synthesis are unable to grow unless they are provided with the end product of that synthesis. For example, the following experiment enables a strain of bacteria in which the synthesis of lysine is blocked to be isolated. A dilute suspension of cells, containing either natural or artificially induced mutants, is spread on a medium which has been solidified by addition of agar and contains all the amino acids likely to be required. After incubation, a large number of colonies of bacteria are visible, each one of which has developed from a single cell. Colonies which have arisen from mutants requiring lysine as a nutrient can be identified as follows. Suppose the colonies appearing on the complete medium form a pattern as in Fig. 3–3(a). Using a piece of sterilized velvet supported on

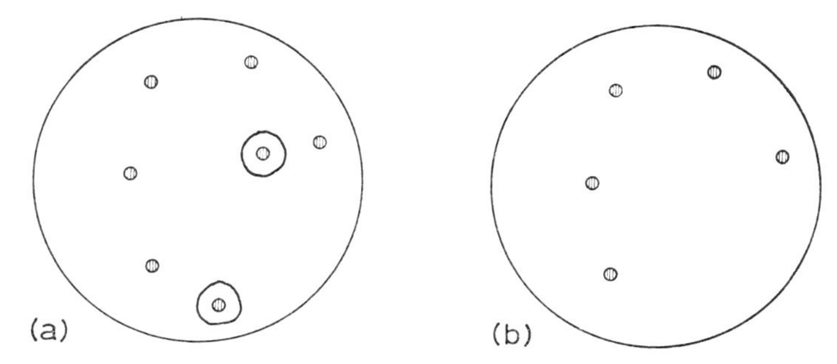

Fig. 3–3 Identification of colonies of mutant bacteria by replica plating. Mutant colonies are ringed.
(**a**) Complete medium.
(**b**) Deficient medium.

a plate of metal, a replica of this pattern of colonies is transferred to another solidified medium from which lysine has been omitted. On incubation, colonies will grow only from cells capable of synthesizing lysine, as shown in Fig. 3–3(b). It can then be concluded that cells in the colonies ringed in Fig. 3–3(a) are unable to synthesize this amino acid. In many micro-organisms, the penultimate link in the biosynthetic chain leading to lysine is α,ϵ-diaminopimelic acid [$HO_2C \cdot CH(NH_2) \cdot (CH_2)_3 \cdot CH(NH_2) \cdot CO_2H$]. If the lysine-requiring mutant were found to grow satisfactorily on a medium containing α,ϵ-diaminopimelic acid, two conclusions could be drawn: first, this compound is an intermediate in the normal formation of lysine in this particular organism; secondly, the block in the biosynthesis of lysine must be earlier than the last reaction of the sequence. Thus it is possible to select mutants having genetic abnormalities appropriate for a chosen biochemical study.

Enzyme reactions may be blocked by artificial means as well as a result of a genetic abnormality and, in passing, it is of interest to compare such abnormalities with the effects of certain drugs. The growth of cultures of the bacterium *Escherichia coli* can be prevented by the addition of sulphanilamide to the medium, without the cells being killed. Such non-growing cultures were found to excrete into the medium the compound aminoimidazole carboxyamide:

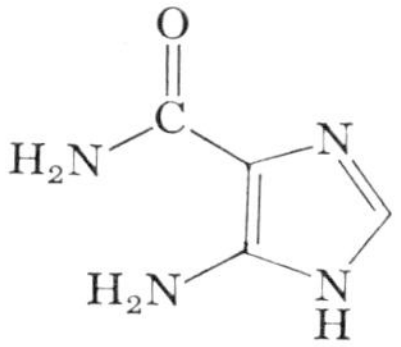

Aminoimidazole carboxyamide

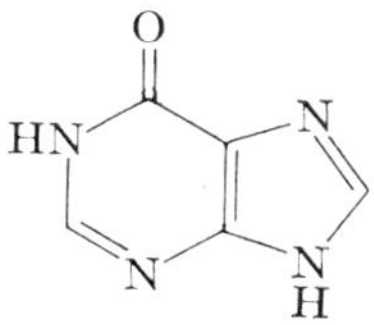

Hypoxanthine

The close relationship of this compound to the purine hypoxanthine can readily be seen. Its accumulation in the culture medium was later ascribed to the inhibition by sulphanilamide of the last stage in the synthesis of the purine ring. Observations such as these can help materially in finding explanations for the mode of action of drugs as well as leading to further insight into the intermediate stages of the chemical processes of the cell.

3·4 The study of biogenesis

Although many enzyme-catalysed reactions are known in detail, a full understanding of cellular chemistry still lies ahead. Much information has accumulated concerning the chemical nature of cell constituents, but finding how they are integrated into the process of life is a complex problem. Fortunately, the use of isotopic tracers allows short cuts to be made in discovering relationships between cell components without the need to find out precisely how interconversions take place.

If certain bacteria are grown in a solution containing acetic acid labelled with the ^{14}C-isotope, it is found that butyric acid which can be isolated from the bacterial cells is radioactive. It is concluded that the butyric acid has been synthesized from the acetic acid supplied in the culture medium and that this is the normal route by which it is formed. This establishes a biogenetic relationship between acetic acid and butyric acid without indicating how the conversion takes place. If acetic acid is used in which only one of the carbon atoms is isotopically labelled, it is found by breaking down the butyric acid in a stepwise manner that the carbon atoms are labelled as shown:

$$2CH_3\cdot{}^*CO_2H \rightarrow CH_3\cdot{}^*CH_2\cdot CH_2\cdot{}^*CO_2H$$

Acetic acid Butyric acid

This demonstrates that the formation of butyric acid involves head-to-tail union between two two-carbon units. By isotopic labelling experiments using whole organisms, the biogenesis of a large number of cell components has been elucidated. In some cases this has led to the discovery of the actual biosynthetic mechanisms and the isolation of the relevant enzymes. However, even in the absence of such detailed information, the study of biogenesis provides a sound basis for an understanding of the inter-relationships of apparently diverse biological materials.

Formic acid serves as a one-carbon unit in the formation of such structures as the purine ring and it is the introduction of this unit which is blocked by sulphanilamide as discussed in § 3·3. Three-carbon units arise in the glycolytic reactions and the reformation of carbohydrates exemplifies their utilization. The five-carbon unit is given the name isoprenoid unit from the compound isoprene:

$$3\,{}^{\circ}CH_3\overset{*}{\cdot}CO_2H \rightarrow HO_2^{*}C\overset{\circ}{\cdot}CH_2\overset{*}{\cdot}C(OH)({}^{\circ}CH_3)\overset{\circ}{\cdot}CH_2\overset{*}{\cdot}CH_2\cdot OH \rightarrow {}^{\circ}CH_2{:}{}^{*}C({}^{\circ}CH_3)\cdot{}^{\circ}CH{:}{}^{*}CH_2$$

Mevalonic acid Isoprene

As indicated in the diagram the five-carbon unit is constructed from three molecules of acetic acid with subsequent loss of carbon dioxide. A very wide variety of cellular materials are built up from isoprenoid units, including the macromolecular hydrocarbons in rubber latex, vitamin A and cholesterol. In the formation of the rubber hydrocarbon, a simple head-to-tail union of isoprenoid units takes place, but in the case of vitamin A cyclization and oxidation also occur:

$$\cdots {}^{\circ}CH_2-{}^{*}C({}^{\circ}CH_3)={}^{\circ}CH-{}^{*}CH_2-{}^{\circ}CH_2-{}^{*}C({}^{\circ}CH_3)={}^{\circ}CH-{}^{*}CH_2\cdots$$

Natural rubber

Vitamin A

The biogenesis of cholesterol, which is referred to in a later chapter, is more complex involving both head-to-tail and head-to-head linkage with concomitant loss of three carbon atoms:

Cholesterol (3-I)

These examples show that biological materials which would defy simple classification on a chemical basis fall readily into a biogenetic group derived from a single type of building unit. It is seen that the living cell creates a large number of different materials with the minimum number of chemical precursors. The study of biogenesis gives glimpses of the general plan of chemical synthesis in the cell and provides a reliable basis for investigations of the detailed processes involved.

Reactions in Cell-free Systems 4

Methods of recognizing biological reactions so far discussed have largely involved whole organisms or at least whole cells. While it is only in the whole organism that chemical reactions proceed undisturbed, the close integration of living processes makes the study of individual reactions difficult. Just as dissection is essential to the discovery of internal morphology, so knowledge of cell chemistry depends on the ability to bring about individual reactions in isolation.

4·1 Cell fractionation

Even the simplest cell exhibits a very complex internal structure when examined with the light microscope and the electron microscope. Figure 4–1 is a composite picture illustrating the various features encountered, all of which are not found in every cell. The distribution of some cell components among the microscopically distinct organelles can be established by using specific chemical stains and by microphotography with ultra-violet light. The determination of the chemical composition of components seen only in electron micrographs presents considerable difficulty. To achieve this it is necessary to separate cell fractions in quantity and to identify as accurately as possible isolated material with structures revealed by the electron microscope.

The first stage in preparing cell fragments is to disrupt the tissue by one of a number of methods such as grinding or homogenization or by subjecting it to high pressures or ultrasonic vibrations. All these processes cause some disintegration of sub-cellular structure but, by centrifugation at different speeds, particles can be prepared from broken cells which are recognizable by microscopy or electron microscopy as entities seen in the whole cell.

The main factors which determine the rate of sedimentation of a suspended particle are the gravitational force acting on it, the size of the particle and its density. In an ordinary laboratory centrifuge, gravitational forces up to $1000 \times g$ can be reached. Machines operating at higher speeds give centrifugal forces up to $5000 \times g$, $20{,}000 \times g$, $100{,}000 \times g$ and above according to the design. Centrifuges giving forces of $100{,}000 \times g$ or above are referred to as ultracentrifuges.

All the distinct parts of the cell shown in Fig. 4–1 cannot be separated. Cell nuclei, the largest and densest particles in suspensions of broken cells, sediment rapidly even in centrifugal fields of a few hundred times gravity, but are difficult to obtain free from chloroplasts, if present, and from debris derived from membranes, etc. Centrifugation in inert liquids of different densities gives some improvement, but complete separation is almost impossible. Mitochondria and lyosomes sediment more slowly and it is

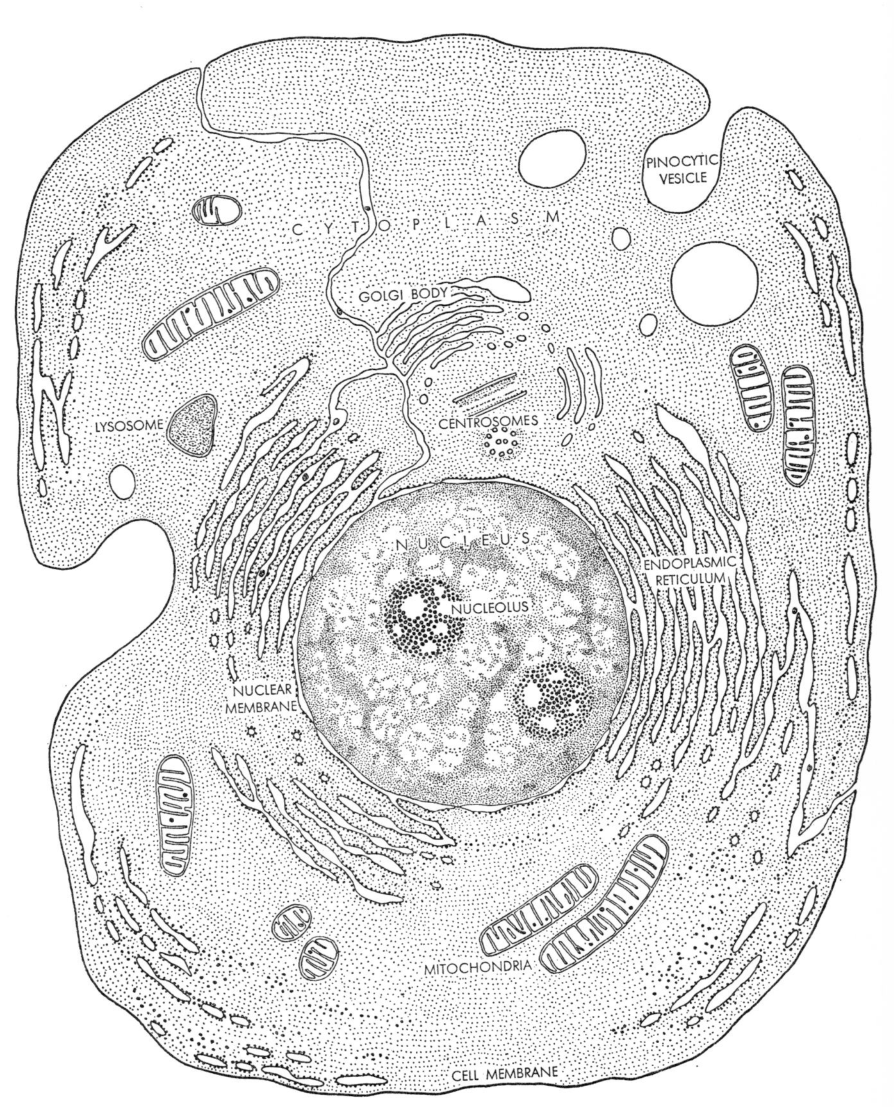

Fig. 4–1 Composite diagram of a typical cell. (Reproduced with permission from Brachet, J. 1961. The Living Cell. *Scientific American*, **205**, 3. Freeman, San Francisco.)

not difficult to choose speeds and times of centrifuging which give a reasonably clean separation. The endoplasmic reticulum is fragmented during homogenization and is sedimented in the ultracentrifuge as the so-called microsome fraction. Parts of the reticulum have a rough appearance because they carry ribosomes, the smallest particles of the cell, some of which are obtained free from reticulum by centrifuging at 100,000 $\times g$. The solution remaining after sedimenting the ribosomes constitutes the supernatant fraction of the cell and contains soluble components but no organized particles.

The efficiency of an experiment in cell fractionation can be assessed in various ways. Microscopic or electron microscopic examination of the material is usually a good guide to the 'cleanness' of a fraction. Experiments with carefully separated fractions have shown that they differ in chemical composition and in the enzymes they contain. The presence of chlorophyll in chloroplasts is an obvious case, but in addition, nuclei are readily identified by the high content of DNA, mitochondria by the presence of such enzymes as glutamic dehydrogenase and lysosomes by the high activity of hydrolytic enzymes such as ribonuclease, phosphatase and protein-hydrolysing (proteolytic) enzymes. Both microsomes and ribosomes contain very high proportions of RNA, but the protein content of the former is higher than the latter due to the presence of the fragmented endoplasmic reticulum.

No isolated particulate fraction of the cell nor the supernatant fraction can fulfil completely its characteristic functions which *in vivo* are so closely integrated. However, by simulation of their natural environment, they can be induced to support to a limited extent their normal chemical reactions. Cell-free preparations in which damage to structures has been reduced to a minimum provide simpler systems for study than whole cells and have made possible considerable advances in knowledge of both cellular components and reactions.

4.2 Enzymes, substrates and co-factors

The separation of sub-cellular fractions considerably reduces the number of chemical reactions which take place in a cell-free system. Nevertheless, each fraction is capable of catalysing many different reactions according to the enzymes which it contains. Each reaction taking place requires a system of some complexity, the general nature of which needs to be appreciated before the complete segregation of biological processes can be discussed. The oxidation of glucose will be taken as an illustration.

If the mould *Penicillium notatum* is provided with a nutrient solution containing glucose, oxygen is consumed and carbon dioxide is produced. However, this gives no indication of the individual reactions taking place. Extracts from the supernatant fraction of the mould bring about only the initial reaction in the combustion of glucose according to the following equation:

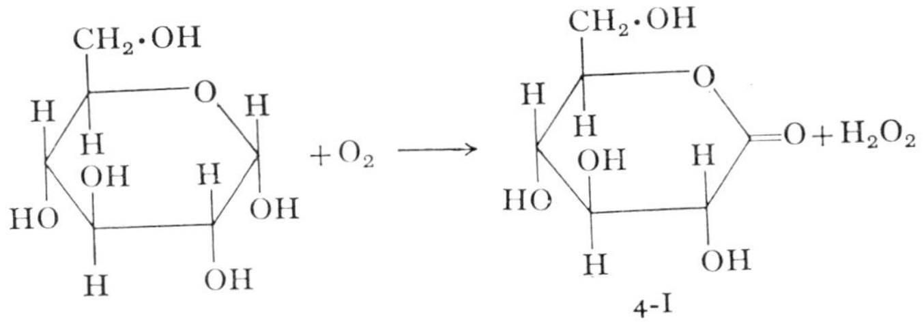

$CH_2{\cdot}OH$ (glucose, pyranose ring: H, H, OH, H, H, OH, HO, H, OH, O) $+ O_2 \longrightarrow$ $CH_2{\cdot}OH$ (ring: H, H, OH, H, HO, H, OH, O) $=O + H_2O_2$

4-I

This reaction is catalysed by an enzyme which specifically acts on glucose as substrate. When first isolated from *Penicillium notatum*, the enzyme was named notatin but is now known as glucose oxidase, or glucose dehydrogenase. The last name most accurately describes the enzyme: it specifies the substrate, it indicates the nature of the reaction catalysed and it has the suffix 'ase' which is used exclusively for enzymes. The action of glucose dehydrogenase on glucose has a number of features of special interest.

The product of the oxidation, glucono-δ-lactone (4-I), is an internal ester of gluconic acid. Gluconic acid readily forms an internal ester on being heated, or in acidic solution. However, under these conditions, ester formation takes place between the carboxyl group and the hydroxyl group located on the fourth carbon of the chain or γ-position. The product is known as glucono-γ-lactone and contains a five-membered ring. Glucono-δ-lactone, in which esterification involves the hydroxyl group on the fifth carbon atom, that is at the δ-position, contains a six-membered ring and could not have been formed from gluconic acid. It must therefore represent the initial reaction product, from which it follows that glucose dehydrogenase reacts specifically with the pyranose form of glucose.

If glucose is incubated with glucose dehydrogenase in presence of ^{18}O-labelled oxygen, it is found that no isotopic oxygen appears in the glucono-δ-lactone, but only in the hydrogen peroxide. Such a result indicates that the oxygen does not take part in a true oxidation of glucose but combines with hydrogen atoms given up by the glucose. It is for this reason, that the name glucose dehydrogenase is preferred to glucose oxidase.

The formation of hydrogen peroxide in the reaction catalysed by this enzyme deserves comment. Before the nature of the enzymic reaction was understood, extracts of the mould were found to have antibacterial action and hopes were held that they might contain a useful drug. It was later realized that the effects were due to the bactericidal properties of the hydrogen peroxide formed, and, since hydrogen peroxide does not show selective toxicity against bacteria only, there was no possible therapeutic use for the enzyme.

Even when highly purified, glucose dehydrogenase from *Penicillium notatum* is yellow in colour, indicating the presence of some structure other than a polypeptide. By precipitation at low pH, the coloured material is removed and a colourless protein is obtained. However, this material is without enzymic activity, showing that the coloured group participates in the catalysis. The coloured compound is called flavin–adenine dinucleotide (FAD) and has the following structure:

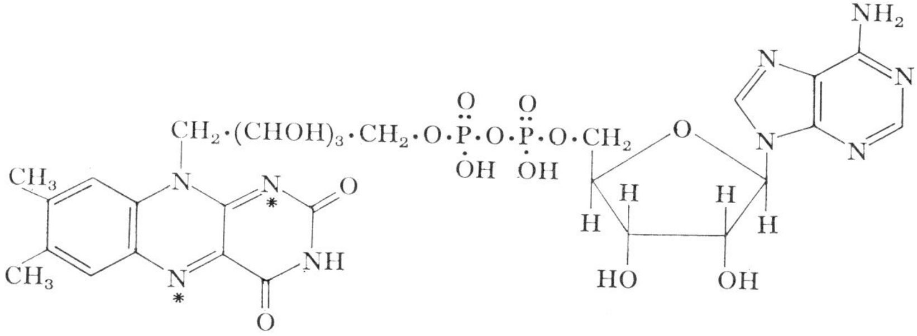

$$CH_2\cdot(CHOH)_3\cdot CH_2\cdot O\cdot P(O)(OH)\cdot O\cdot P(O)(OH)\cdot O\cdot CH_2$$

Flavin–adenine dinucleotide

Hydrogen atoms removed from the molecule of glucose become attached to the nitrogen atoms marked in FAD before being passed on to molecular oxygen in the formation of hydrogen peroxide.

It is seen that the initial step in the combustion of glucose by *Penicillium notatum* requires a complex system for its fulfilment and separation of the various components has enabled a detailed picture of the reaction to be built up. Many other enzymic reactions have been studied in this way. Most are found to resemble glucose dehydrogenase in general pattern in requiring a non-protein component, the co-factor, in addition to the catalytic enzyme protein specific for the substrate. Some enzymes, notably those concerned with hydrolyses, do not require a co-factor. Where a co-factor is involved, its character is largely determined by the nature of the reaction catalysed. FAD serves as a hydrogen carrier in a number of dehydrogenations. In others, transfer of hydrogen is effected by a closely related compound, flavin mononucleotide (FMN):

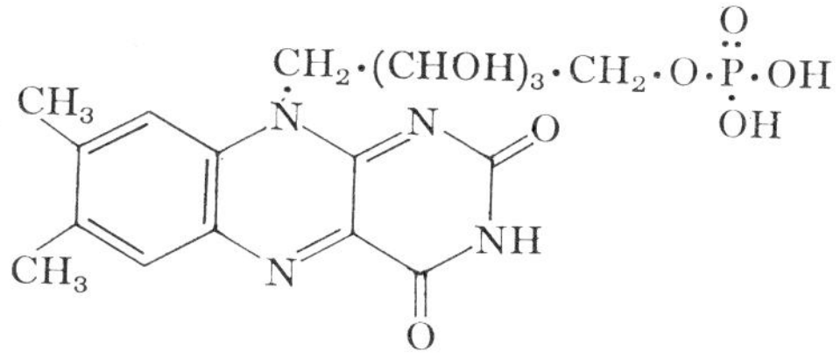

$$CH_2\cdot(CHOH)_3\cdot CH_2\cdot O\cdot P(O)(OH)\cdot OH$$

Flavin mononucleotide

Other examples include nicotinamide adenine dinucleotide (NAD) (4-II, R=H) and nicotinamide adenine dinucleotide phosphate (NADP) (4-II, $R=OPO_3H_2$):

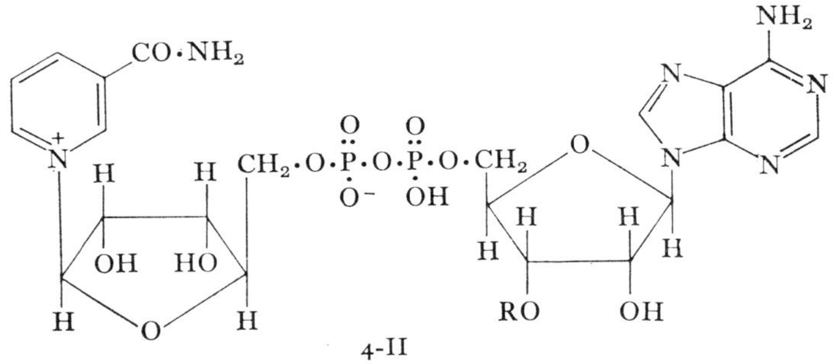

4-II

Both these co-factors accept hydrogen atoms from the substrate and are thereby reduced respectively to dihydronicotinamide–adenine dinucleotide ($NADH_2$) and dihydronicotinamide–adenine dinucleotide phosphate ($NADPH_2$) containing in each case a reduced nicotinamide ring. ATP is a co-factor of a different type altogether, being concerned in reactions in which phosphate residues are transferred. The rôles of NADP and ATP as co-factors are illustrated by an alternative series of reactions involved in the oxidation of glucose.

Penicillium notatum is exceptional in its mode of utilization of glucose. More frequently, oxidation of glucose is preceded by its conversion into glucose 6-phosphate. This route, known as the direct oxidative pathway of glucose metabolism was discovered in red blood cells and involves the following reactions as the first two steps:

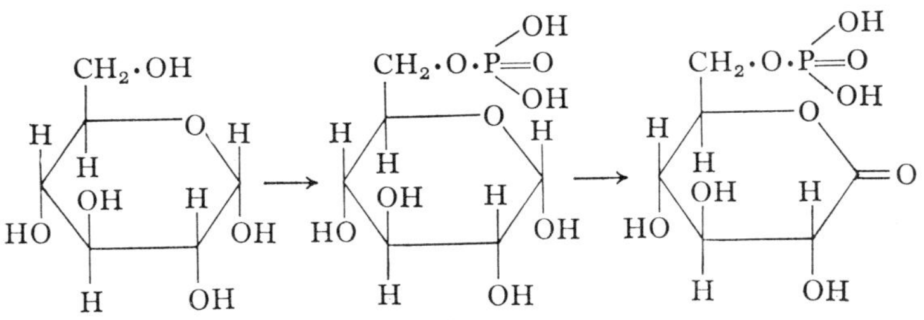

The initial stage is catalysed by the enzyme glucokinase in presence of ATP as co-factor, the latter being converted into ADP in the process. In the second reaction, the enzyme glucose 6-phosphate dehydrogenase catalyses the transfer of hydrogen from the substrate to NADP, which is thereby converted into $NADPH_2$. These two enzymes differ from glucose dehydrogenase in one important respect. Whereas FAD is bound to the protein in glucose dehydrogenase and is only removed by chemical means, the co-factors in glucokinase and glucose 6-phosphate dehydrogenase are not firmly attached to the enzymes and are removed during isolation. A further feature of enzymes such as glucokinase using ATP as co-factor is that the association of protein and co-factor, necessary for catalytic activity, requires the presence of magnesium ions.

An enzyme catalysed reaction is the simplest system which truly represents a chemical reaction taking place in a cell. From the examples discussed above, it is seen that a number of components may be required to effect a single transformation of substrate into product. Only by careful separation and reconstruction can the detailed mechanism of such reactions be elucidated. Many such enzyme systems have been successfully taken to pieces and put together again. Others have so far resisted dissection in this way and much effort is required to devise new methods of fractionation which will allow the more delicate chemical assemblies to be taken apart without destroying the essential environment required for the operation of the biological processes concerned.

4.3 Separating large and small molecules

It will now be appreciated that the operational units of cell chemistry contain both large and small molecules, and their full characterization requires the separation and identification of all components. The rapid advances in biochemistry over the last two decades have depended very largely on the development of methods of fractionation. Space does not permit the discussion of all available techniques, but examples have been chosen to illustrate the principal methods.

Ionized molecules are most frequently separated from each other by electrophoresis or by ion-exchange chromatography. The former depends on the movement of ions in an electric field. Movement due to convection is prevented by supporting the solution on filter paper or by converting it into a gel. The rate of movement of an ion towards one or other electrode is determined by the net charge on the ion and the size of the molecule. The behaviour of amino acids on electrophoresis depends critically on the pH of the solution because of the existence of different ionized forms (see Chapter 2). At a pH known as the isoelectric point for a particular amino acid, the net charge on the molecule will be zero and no movement will take place on application of an electric field. By varying the pH of the solution, conditions can be found in which different amino acids move at different rates due principally to slight differences in the net charge on the molecules. The technique is equally applicable to other ionized molecules and is used to a considerable extent in fractionation of nucleotides, carboxylic acids and many other types of cell component. Of the factors affecting the rate of migration of an ion, net charge is the most important. However, the effect of molecular size becomes appreciable in the electrophoresis of proteins and nucleic acids which move more slowly than their constituent amino acid or nucleotide units.

Ion-exchange chromatography depends on the combination of a dissolved ion with an insoluble ion of opposite polarity which is usually in the form of a resin particle. For the separation of amino acids, a polystyrene resin in bead form carrying sulphonic acid residues is most frequently used. A solution of amino acids at low pH is percolated through a column of the resin and since under these conditions the basic groups are positively charged, the amino acids are adsorbed by salt formation with the negatively charged sulphonate ions:

$$\text{Resin—}\overset{-}{\text{SO}_3}\overset{+}{\text{H}} + \overset{-}{\text{Cl}}\text{H}_3\overset{+}{\text{N}}\cdot\text{CHR}\cdot\text{CO}_2\text{H} \rightarrow \text{Resin—}\overset{-}{\text{SO}_3}\text{H}_3\overset{+}{\text{N}}\cdot\text{CHR}\cdot\text{CO}_2\text{H} + \overset{+}{\text{H}}\overset{-}{\text{Cl}}$$

On washing the resin with solutions of increasing pH, the amino acids are released according to the following equation:

$$\text{Resin—}\overset{-}{\text{SO}_3}\text{H}_3\overset{+}{\text{N}}\cdot\text{CHR}\cdot\text{CO}_2\text{H} + \overset{+}{\text{Na}}\overset{-}{\text{OH}} \rightarrow \text{Resin—}\overset{-}{\text{SO}_3}\overset{+}{\text{Na}} + \text{H}_3\overset{+}{\text{N}}\cdot\text{CHR}\cdot\overset{-}{\text{CO}_2} + \text{H}_2\text{O}$$

The most weakly basic amino acid is removed from the resin first and others are eluted successively in order of increasing basic strength. This method of fractionation is almost universally used for quantitative analysis of mixtures of amino acids and the technique has been so standardized that amino acids can be identified by the sequence in which they are eluted. Similar methods are also used for separation of nucleotides, but in this case, resins carrying basic residues are employed which adsorb the nucleotides at high pH by virtue of the dissociated phosphate groups. The nucleotides are eluted in sequence at gradually decreasing pH values. Ion-exchange chromatography has been used also for proteins and nucleic acids, but with some resins adsorption is poor because large molecules are unable to penetrate the molecular lattice of the resin.

Many biological molecules do not carry readily ionizable groups and consequently cannot be separated by either of the methods so far discussed. Carbohydrates and other hydroxylic compounds are most readily fractionated by paper chromatography which makes use of differences in partition coefficient between partially miscible liquids. This technique is of wide application and experimental details are fully described by FEINBERG and SMITH (see Further Reading).

Fatty acids, although they contain ionizable groups, are more readily fractionated by adsorption on to inert solids such as silica or kieselguhr. The molecules may be adsorbed from solution but far greater resolution is obtained by adsorption from the vapour phase. Long-chain fatty acids which formerly resisted fractionation can now be separated by vapour-phase chromatography.

Most techniques of fractionation so far described find their major applications in the resolution of mixtures of small molecules. One of the most useful operations in studying cell components is the segregation of large from small molecules. While a gross separation can be achieved by dialysis, much more effective methods are now available. The processes involve chromatography on colloidal preparations of modified polysaccharides. Such materials can be prepared with varying degrees of linking between chains, and the colloidal particles consist of open or dense networks according to the degree of cross-linking. If a solution containing both large and small molecules is percolated through a column of material with a low degree of cross-linking, all the molecules are able to penetrate the particles and, being adsorbed, are retained by the column. However, if a polysaccharide having a high degree of cross-linking is used, the dense network allows only small molecules to penetrate the colloidal particles; large molecules cannot, and are not retained by the column. These two situations are illustrated in Fig. 4–2. This technique is known as molecular-

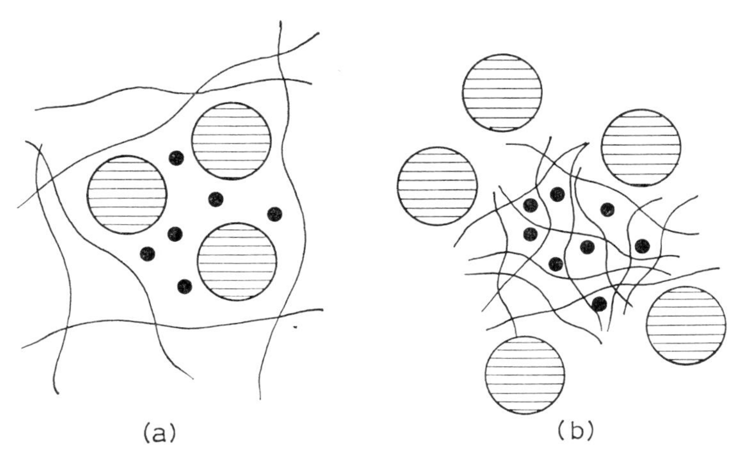

Fig. 4–2 (**a**) Retention of large and small molecules by colloidal particle of polysaccharide with low degree of cross-linking. (**b**) Exclusion of large molecules from particle of highly cross-linked polymer.

sieve chromatography, or gel filtration. It can be used not only for separating large from small molecules, but also, by calibration of the porosity of the gel with molecules of known size, can provide a useful estimate of molecular weight.

One of the most important processes of separation in biochemistry is the fractionation of mixtures of proteins, since this is fundamental to the preparation of enzymes specific for the catalysis of individual chemical reactions. The general structural similarities between proteins makes it difficult to distinguish between them. Some separation can be achieved by electrophoresis and ion-exchange chromatography by virtue of differences in the proportions of basic and acidic amino acid residues. However, these techniques are generally not as efficient for the separation of large molecules as for small. One exception must be made. By using a cross-linked gel as support for the solution, instead of filter paper, proteins may be separated with very high resolution employing simultaneously the principles of electrophoresis and gel filtration. For the preparative separation of proteins selective precipitation is usually the method of choice.

The solubilities of proteins in aqueous solution vary considerably according to the presence or absence of other solutes. Figure 4–3 illustrates the solubilities of different proteins of the blood in presence of

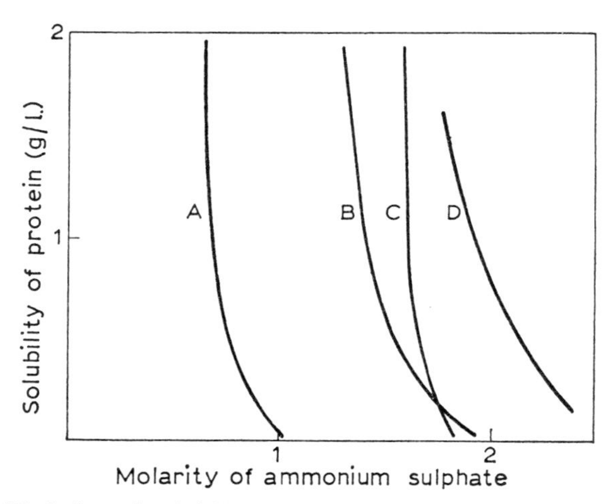

Fig. 4–3 Variation of solubilities of blood proteins with concentration of ammonium sulphate. A, fibrinogen; B, haemoglobin; C, pseudoglobulin; D, serum albumin.

varying concentrations of ammonium sulphate and it is clear that by gradually adding ammonium sulphate, the components could be precipitated in succession. This process of 'salting out' results from the solvation of the salt ions, which has the effect of leaving insufficient water molecules free to provide an 'atmosphere' of solvent round the protein molecules. In this condition, the positive and negative charges of the basic and acidic amino acid residues in the proteins exert a force of attraction between adjacent protein molecules resulting in aggregation and precipitation.

The force of attraction between electrically charged particles is dependent not only on the size of the positive and negative charges and the distance between them, but also on the dielectric constant of the intervening medium. The force of attraction is high for a medium of low dielectric constant. The value for water at 25° is 78 while that of alcohol at the same temperature is 25. Thus by adding alcohol to an aqueous solution of a protein, the dielectric constant of the solvent is reduced. The force of attraction between the protein molecules is thereby raised and the increased tendency towards aggregation results in a reduction in solubility. This method gives very high selectivity in the precipitation of proteins from mixtures. It has the disadvantage that alcohol brings about denaturation of proteins unless the temperature is kept low and for this reason other methods are more frequently employed.

The preparation of ovalbumin from egg white illustrates the method most often used and is straightforward to carry out. If an equal volume of saturated ammonium sulphate solution is added, with stirring, to egg white in a beaker, much of the protein is precipitated and may be filtered off using a large fluted filter paper. Ovalbumin remains in solution in 50% saturated ammonium sulphate but is precipitated from the filtrate by first adding more saturated ammonium sulphate solution until there is a slight permanent turbidity (approximately 3–5 ml. for each 100 ml. of filtrate will be needed). Then 0·2 N-H_2SO_4 is added to bring the pH to 4·6, the solution being vigorously stirred during the addition. The ovalbumin is precipitated making the solution turbid and on standing for a few days crystals of the protein appear and can be seen as fine needles under the microscope.

The Supply and Use of Energy 5

As mentioned earlier, one of the central requirements for the maintenance of life is a continuous supply of energy. It would be logical to consider first the source of energy and then to examine the ways in which it is stored and subsequently made to serve the purposes of the living cell. An appreciation of the energetics of cell chemistry requires an insight into the interrelationship of many different chemical events and it is helpful to deal first with the ways in which the energy contained in carbon compounds is converted into a form readily available for use.

5.1 Anaerobic metabolism

Although glucose is only one of many different carbon compounds oxidized by cells for the generation of energy, its metabolism illustrates most clearly the processes involved. We have already seen how glucose or its phosphoric ester is oxidized to derivatives of gluconic acid. Oxidation is important in the provision of energy only in so far as it is accompanied by the fission of carbon–carbon bonds. In this respect the oxidations of glucose or glucose 6-phosphate discussed in Chapter 4 are not intrinsically significant. Their relevance to energy production will become clear later. In the majority of cells, glucose is fragmented to smaller molecules before oxidation takes place, thus making the energy derived from the fission of bonds immediately available.

The splitting of glucose into smaller fragments, known as glycolysis, takes place by similar pathways in most cells, but has been studied in greatest detail in yeast and muscle. The reactions occurring in these two tissues are identical up to a point and are shown in Fig. 5–1.

As shown, the sum total of these reactions is the dehydrogenation of one molecule of glucose at the expense of NAD, an apparently meagre result for such a complicated series of steps. The significance of the sequence of reactions becomes apparent when the participation of ATP and ADP is considered. Each six-carbon unit, namely glucose, requires two molecules of ATP for the formation of fructose 1,6-diphosphate. The latter molecule yields two three-carbon units of glyceraldehyde 3-phosphate, each of which combines with H_3PO_4 during its dehydrogenation. Thus for each unit of six carbon atoms, two molecules of ATP are recovered during the formation of 3-phosphoglyceric acid, and there is subsequently a net formation of two molecules of ATP by virtue of the last reaction in the sequence. As has already been explained in Chapter 2, ATP is a very reactive compound, the hydrolysis of which liberates a considerable amount of energy. How this energy is harnessed by the cell for useful purposes will be

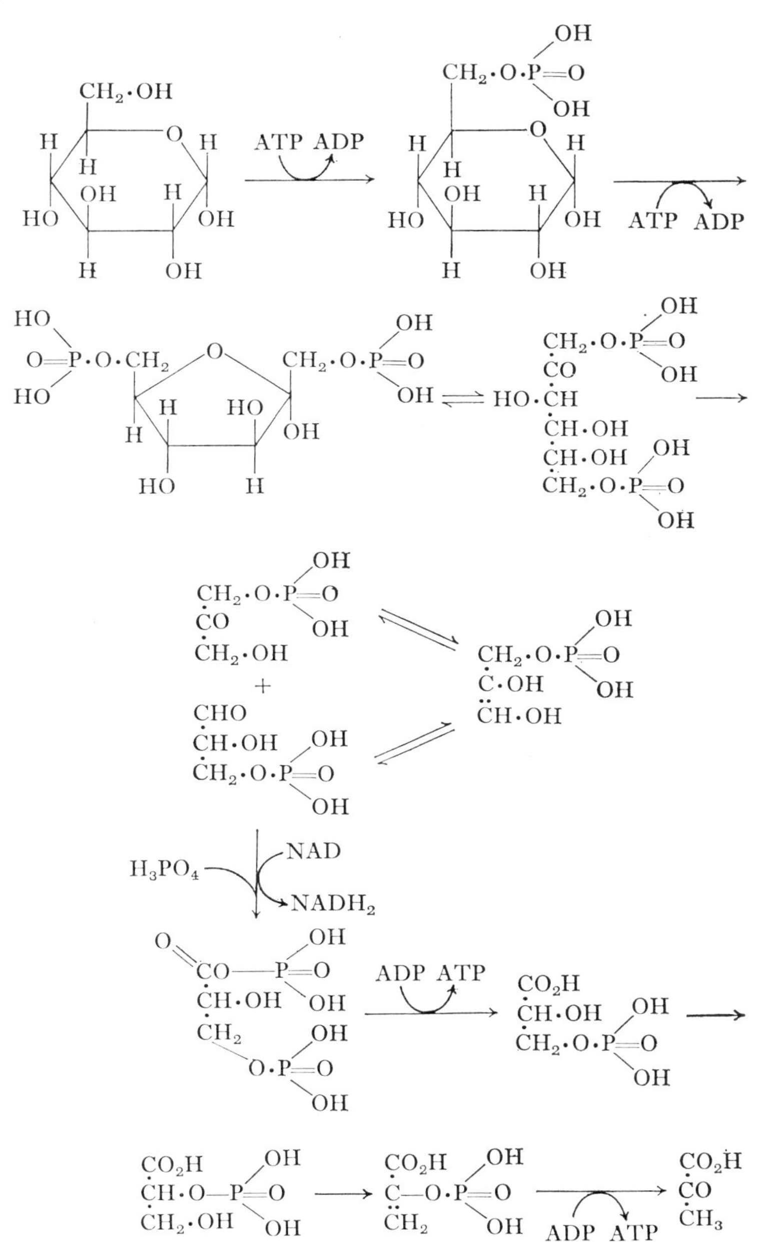

Fig. 5–1 The reactions of glycolysis.

discussed in a later section, but it is clear that the process of glycolysis provides the cell with a store of energy in the form of two molecules of ATP for each molecule of glucose used. ATP is not the only energy-yielding molecule found in cells. 1,3-Diphosphoglyceric acid is another example. However, because ATP participates in so many cellular reactions, it serves as a common currency for the transfer and utilization of energy and the usefulness of a metabolic sequence of reactions can be measured in terms of its yield of ATP. Viewed in this light, the glycolytic sequence of reactions is a potentially useful one since it involves the net transfer of a phosphate residue from substrates to ADP. This happens at the expense of the overall oxidation of glucose by NAD and without the participation of molecular oxygen. In the presence of oxygen, $NADH_2$ is re-converted to NAD by reactions to be discussed later. In muscle tissue in the absence of oxygen the supply of NAD is replenished by the following reaction of $NADH_2$ with pyruvic acid, which is catalysed by the enzyme lactic dehydrogenase:

$$NADH_2 + CH_3 \cdot CO \cdot CO_2H \rightarrow NAD + CH_3 \cdot CH(OH) \cdot CO_2H$$

The regeneration of NAD allows glycolysis to continue with the further production of ATP. Thus a muscle which is working hard and requiring energy at a faster rate than it can receive oxygen from the blood, can produce ATP by glycolysis with the resultant accumulation of lactic acid. This process does not require oxygen and is known as anaerobic glycolysis. The overall reaction is:

$$C_6H_{12}O_6 \rightarrow 2CH_3 \cdot CH(OH) \cdot CO_2H$$

In yeast, pyruvic acid is formed from glucose by precisely the same reactions as in muscle. However, the anaerobic regeneration of NAD takes place by decarboxylation of pyruvic acid and the reduction of the acetaldehyde so formed to yield ethanol under the influence of alcohol dehydrogenase:

$$CH_3 \cdot CO \cdot CO_2H \rightarrow CH_3 \cdot CHO + CO_2$$

$$CH_3 \cdot CHO + NADH_2 \rightarrow CH_3 \cdot CH_2 \cdot OH + NAD$$

Anaerobic glycolysis in muscle and yeast enables ATP production to be effected in the absence of molecular oxygen. It cannot maintain muscular activity or the growth of yeast indefinitely because of the toxicity of high concentrations of lactic acid and ethanol respectively. It does, however, act as a device for ensuring continuity in the supply of ATP during a temporary shortage of oxygen. Pyruvic acid is also formed during the metabolism of 6-phosphogluconic acid, but since the initial oxidation of glucose 6-phosphate to the lactone of 6-phosphogluconic acid utilizes NADP rather than NAD (see Chapter 4), reduction of pyruvic acid cannot support the continued oxidation of glucose 6-phosphate. The direct oxidative pathway of glucose utilization cannot therefore operate under anaerobic conditions.

Anaerobic glycolysis is an example of a type of metabolic sequence of reactions known as a fermentation. Many other fermentations are known and all are characterized by an oxidative and a reductive stage. Various end products other than lactic acid or ethanol are formed by different organisms but in most cases pyruvic acid serves as the key intermediate linking the two halves of the process. A number of useful products such as butanol and acetone are commonly manufactured by fermentation. Most organisms utilize foodstuffs either by fermentation or by respiration according to the availability of oxygen. However, some micro-organisms, the so-called strict anaerobes, only use fermentation processes and are killed by molecular oxygen. Such organisms are relatively few in number, the majority depending ultimately on respiration.

5.2 The chemistry of respiration

Anaerobic glycolysis results in the fission of only one carbon–carbon bond in each molecule of glucose. Respiration, which involves the complete break-up of the molecule to give carbon dioxide, is much more efficient in making energy available in the form of ATP. Pyruvic acid is a key compound in the complete combustion of glucose as well as in anaerobic glycolysis and this enables cells to switch readily from anaerobic to aerobic metabolism. A muscle which has been working anerobically can, under aerobic conditions, dispose of the accumulated lactic acid by re-conversion to pyruvic acid under the influence of lactic dehydrogenase. The pyruvic acid is then completely oxidized in the process of respiration. The mechanism of oxidation differs from one organism to another and from one tissue to another, but one reaction sequence stands out in importance from the rest. Once again, the reactions have been studied most extensively in muscle.

Slices of fresh muscle tissue when incubated in a Warburg respirometer continue to consume oxygen and produce carbon dioxide for a limited time. The rate of respiration gradually falls but can be restored by adding one or more of a number of compounds, such as succinic acid, fumaric acid, malic acid, oxaloacetic acid, citric acid or oxoglutaric acid. After the addition of even small quantities of *one* of these compounds, oxygen consumption and carbon dioxide production continues to an extent far greater than that corresponding to the oxidation of the compound added. That is, these compounds behave catalytically in their effect on respiration. In 1937, H. A. Krebs put forward an explanation in which the compounds enumerated above are alternately destroyed and re-formed in a cyclic scheme of reactions, which accounts for the small quantities required. Enzymes which catalyse all the postulated interconversions have been separated from muscle tissue and the cyclic reaction pathway, usually referred to as the citric acid cycle, is shown in Fig. 5–2. The three salient features of this cyclic process are:

1. The participation of pyruvic acid in the formation of citric acid from oxaloacetic acid.

2. The formation of three molecules of carbon dioxide in the complete cycle of reactions.

3. The consumption of two molecules of NAD and one of FAD. These facts imply that provided a catalytic quantity of any one of the intermediates in the cycle is present, together with the necessary enzymes,

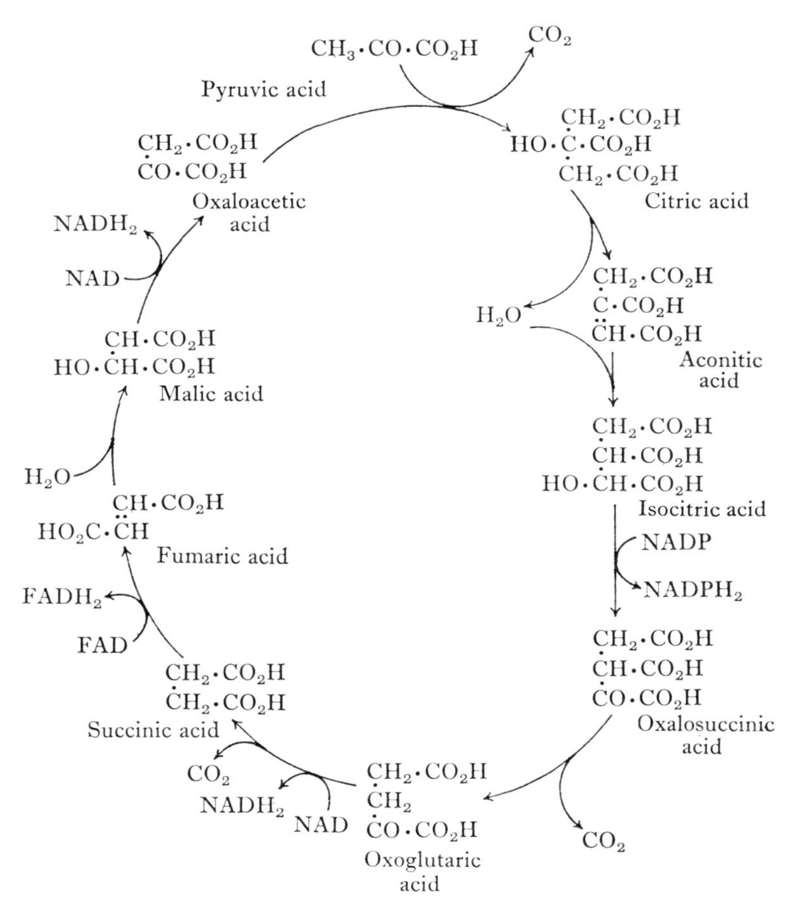

Fig. 5–2 Reactions of the citric acid cycle.

pyruvic acid can be oxidized to carbon dioxide in an amount related directly to the quantities of NAD and FAD available. Since pyruvic acid is completely combusted in this process, its reduction cannot bring about the reoxidation of reduced co-factors. This is achieved ultimately through oxidations involving molecular oxygen. The part played by oxygen in respiration is largely concerned with regeneration of co-factors by a sequence called the respiratory chain of reactions. Moreover, the reactions of the respiratory chain are closely associated with the formation of ATP from ADP by a process known as oxidative phosphorylation.

Experiments with isolated mitochondria have shown that the enzymes associated with the citric acid cycle, the respiratory chain and with oxidative phosphorylation are located in these particles. They are firmly bound to the mitochondria, constituting a highly organized unit for the oxidation of pyruvic acid and the generation of ATP. It is found that the various enzymes concerned are present in fixed proportions thus ensuring smooth running of the production line. Disruption of the mitochondria breaks the continuity of the process and many of the enzymes cannot be obtained in solution separate from the organized particles.

The chemical reactions taking place in mitochondria have proved most difficult to disentangle and are still not fully understood, because of the difficulty of separating individual components. $NADH_2$ is not oxidized directly by molecular oxygen but is reconverted to NAD by an enzyme, $NADH_2$ dehydrogenase, which carries a flavin co-factor. The reduced flavin, which is produced at the expense of the oxidation of $NADH_2$, is re-oxidized by a group of metal-containing proteins known as the cytochromes. These form a complex group of enzymes, of which cytochrome-*a*

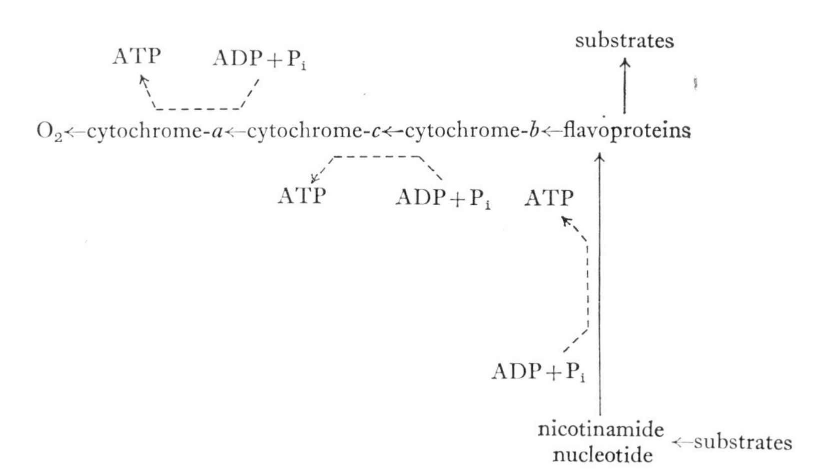

Fig. 5–3 The main pathway of respiratory oxidation. Heavy arrows indicate that the compound at the tail of the arrow is oxidized by that at the head; they also show the direction of flow of electrons in the chain. Dotted arrows indicate the sites in the chain at which the formation of ATP is coupled with the flow of electrons.
P_i = orthophosphate.

(formerly called cytochrome oxidase) is readily oxidized by molecular oxygen. In the mitochondrion this process is always associated with the oxidation by cytochrome-*a* of a further component of the system, namely cytochrome-*c*, the only one which has been isolated from mitochrondria in pure form. In its oxidized form, it is capable of effecting the oxidation of a third cytochrome, namely cytochrome-*b*, which in turn oxidizes the flavin-containing $NADH_2$ dehydrogenase. This sequence of reactions, which is summarized in Fig. 5–3, does not include all the interconversions involving oxidation of co-factors, but constitutes the main pathway of respiration. The process may seem unnecessarily complex, but, as shown in Fig. 5–3, formation of ATP occurs at three stages in the oxidation and if co-factors were regenerated by direct reaction with oxygen, the crucial production of ATP would be by-passed. Precisely how the phosphorylation of ADP is linked to the various oxidations is not known, but about forty molecules of ATP are produced from all the oxidations involved in the conversion of one molecule of glucose into carbon dioxide and water. This yield of ATP from respiratory reactions compares very favourably with the meagre two molecules of ATP per molecule of glucose undergoing anaerobic glycolysis.

The chemistry of respiration has been discussed in terms of the utilization of glucose for the provision of energy in the form of ATP. Since glucose is not the sole item of food, it is necessary to enquire how general are the respiratory reactions in the utilization of nutrients. In fact, the citric acid cycle and the respiratory chain reactions are capable of accomplishing the oxidation of nearly all of the combustible components of food. Pyruvic acid is formed from foodstuffs other than glucose, for example from the amino acid alanine by oxidation and deamination:

$$\begin{array}{l} CH_3\cdot CH\cdot CO_2H + \tfrac{1}{2}O_2 \rightarrow CH_3\cdot CO\cdot CO_2H + NH_3 \\ \quad\quad\;\; \dot{N}H_2 \end{array}$$

The formation of citric acid from pyruvic acid takes place in two stages, the first of which involves decarboxylation to a derivative of acetic acid in combination with the cofactor coenzyme A:

$$CH_3\cdot CO\cdot CO_2H + CoA \rightarrow CO_2 + CH_3\cdot CO\cdot CoA$$

The second stage is the reaction of acetyl coenzyme A with oxalo-acetic acid:

$$\begin{array}{lcl} & & CH_2\cdot CO_2H \\ CH_3\cdot CO\cdot CoA + CO\cdot CO_2H & \rightarrow & HO\cdot\dot{C}\cdot CO_2H \quad + CoA \\ \quad\quad\quad\quad\quad\quad\;\; \dot{C}H_2\cdot CO_2H & & \dot{C}H_2\cdot CO_2H \end{array}$$

Thus acetyl coenzyme A is the crucial compound in the operation of the citric acid cycle. It is formed directly from acetic acid in presence of a specific enzyme, but also arises from higher fatty acids by successive oxidations at the carbon atom in the position in the chain next but one to the carboxyl group. Coenzyme A also participates in this process:

$$R\cdot CH_2\cdot CH_2\cdot CH_2\cdot CH_2\cdot CO\cdot CoA \rightarrow$$

$$R\cdot CH_2\cdot CH_2\cdot CO\cdot CH_2\cdot CO\cdot CoA \xrightarrow{CoA}$$
$$R\cdot CH_2\cdot CH_2\cdot CO\cdot CoA + CH_3\cdot CO\cdot CoA$$

In addition to these alternative sources of material for the formation of citric acid, a number of foodstuffs give rise to other participants in the citric acid cycle. Thus oxidative deamination of glutamic acid yields α-oxoglutaric acid, which on further metabolism yields reduced cofactors which are fed into the respiratory chain.

To summarize, the citric acid cycle, although not the exclusive reaction pathway for the oxidation of cell nutrients, is the major one in most tissues and is capable of oxidizing not only carbohydrates present in food, but also the degradation products of fats and proteins, with the simultaneous production of reduced NAD and FAD. The respiratory chain is able to reoxidize these materials from whatever source and, although this is not the sole source of ATP, the production of this compound which accompanies the oxidation of co-factors represents the major source of energy in most cells.

5·3 Energy from the sun

In the first two sections of this chapter, chemical reactions have been discussed by which energy derived from food is made accessible to the cell. However, continuity in the provision of energy presupposes an unlimited supply of oxidizable carbon compounds. Clearly if the energy required for the synthesis of such materials was derived only from the fission of other carbon compounds, the system would quickly come to an end. The majority of organisms, which have an absolute requirement for organic foodstuffs, are ultimately dependent on photosynthetic organisms which harness the energy of the sun's radiation to the synthesis of carbon to carbon bonds. Photosynthesis is carried out by all green plants and in addition certain micro-organisms such as *Euglena*, *Chlorella* and some bacteria. The overall result of the process in higher plants is the reduction of carbon dioxide to hexose with the production of oxygen:

$$6CO_2 + 6H_2O \rightarrow C_6H_{12}O_6 + 6O_2$$

In photosynthetic bacteria carbon dioxide is reduced but no molecular oxygen is formed. Photosynthesis takes place in specialized particles, the chloroplasts, in higher plants, or the chromatophores in photosynthetic bacteria, which all contain the pigment chlorophyll, together with other pigments such as β-carotene. Chlorophyll occupies the central role in absorbing red light (wave-length around 700 mμ) and making the energy associated with it available for driving the photosynthetic chemical reaction. Other pigments present absorb radiation of shorter wave-lengths and, in some way not understood, transfer the absorbed energy to the chlorophyll molecule, thus extending the effective range of wave-length.

Chlorophyll consists of four heterocyclic rings linked round a central magnesium atom to form a large cyclic structure containing alternating single and double bonds. In such systems, the valency electrons associated with the double bonds become more mobile and the absorption of light results in the ejection of an electron from the chlorophyll molecule. This electron can be taken up by other molecules which are thereby reduced. The acceptance of an electron is equivalent in chemical terms to reduction, as can be seen by considering the conversion of the ferric ion to the ferrous ion:

$$Fe^{3+} + e \rightleftharpoons Fe^{2+}$$

The availability of an electron resulting from the absorption of light by chlorophyll makes possible the reduction of carbon dioxide. Conversely, the loss of the electron by chlorophyll converts the molecule into an oxidizing agent which is ultimately responsible for the oxidation of water to molecular oxygen. The intermediate stages leading to the formation of glucose and oxygen are complex. Illumination of chloroplasts in the absence of carbon dioxide results in the evolution of oxygen, which is therefore clearly independent of the fixation of carbon. In fact, by using ^{18}O-labelled water, it can be shown that the oxygen gas produced in photosynthesis is derived from water and not at all from carbon dioxide. Thus photosynthesis is more accurately described in the following equation in which the fate of the labelled oxygen atoms is indicated:

$$6CO_2 + 6H_2{}^{18}O \rightarrow C_6H_{12}O_6 + 6{}^{18}O_2$$

This formulation of the process emphasizes what is believed to be the primary reaction of the photo-activated chlorophyll molecule, namely the splitting of a molecule of water. However, before discussing the fission of the water molecule and its consequences, it is desirable to consider the fate of carbon dioxide and its conversion into glucose.

If leaves are illuminated in an atmosphere containing ^{14}C-labelled carbon dioxide, radioactivity is detectable in a number of constituents of the leaf even after a few seconds, and this fixation of carbon proceeds for a short time after illumination has stopped. Fractionation shows that, after a short pulse of illumination, the highest radioactivity is present in the compound 3-phosphoglyceric acid. This compound participates in the reactions of the glycolytic pathway, and since these reactions are reversible, it is easy to envisage how glucose is formed from it. However, the route whereby carbon dioxide enters the molecule of phosphoglyceric acid is a devious one as shown:

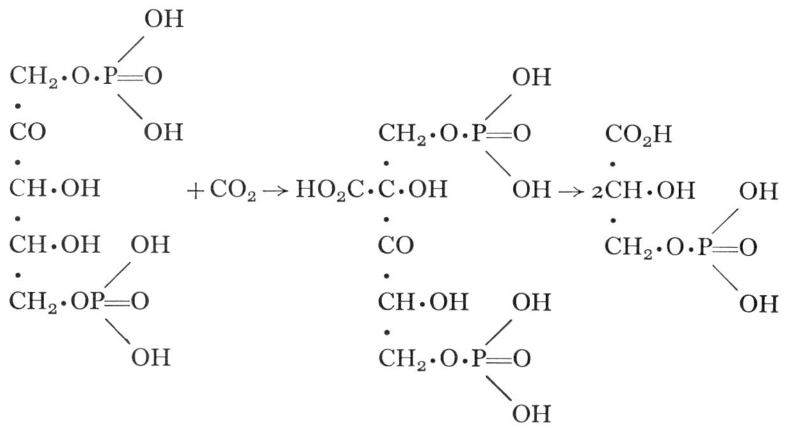

```
          OH
         /
CH2·O·P=O                                       OH
·        \                                     /
CO        OH                          CH2·O·P=O        CO2H
·                                     ·        \       ·
CH·OH          + CO2 → HO2C·C·OH               OH → 2CH·OH    OH
·                                     ·                ·      /
CH·OH    OH                           CO               CH2·O·P=O
·       /                             ·                       \
CH2·OP=O                              CH·OH     OH             OH
        \                             ·        /
         OH                           CH2·O·P=O
                                               \
                                                OH
```

Ribulose 1,5-diphosphate 3-Phosphoglyceric acid

Ribulose 1,5-diphosphate arises by phosphorylation of ribulose 5-phosphate which in turn is formed from 6-phosphogluconic acid. For the photosynthetic fixation of carbon dioxide to proceed continuously, ribulose 1,5-diphosphate must be regenerated and this takes place at the expense of part of the phosphoglyceric acid initially formed. In fact for each six molecules of carbon dioxide reacting with ribulose 1,5-diphosphate, one molecule of hexose is formed together with six molecules of ribulose 5-phosphate:

$$
\begin{array}{lrcl}
 & 6C_5 + 6C_1 & \rightarrow & 12C_3 \\
 & 2C_3 & \rightarrow & C_6 \\
 & 10C_3 & \rightarrow & 6C_5 \\
\hline
\text{Sum:} & 6C_1 & \rightarrow & C_6 \\
\hline
\end{array}
$$

ATP is required for the regeneration of ribulose 1,5-diphosphate from ribulose 5-phosphate, and also in the conversion of 3-phosphoglyceric acid into hexose. In the glycolytic sequence, ATP is produced during the conversion of glyceraldehyde 3-phosphate into 3-phosphoglyceric acid (see Chapter 5.1) and consequently ATP is required for the reversal of this process. The reduction of 3-phosphoglyceric acid in plants also requires $NADPH_2$ and it is in the provision of this and of ATP that the photochemical part of photosynthesis, the so-called light reaction, is concerned. We therefore return to the primary reaction of photosynthesis, namely the photolysis of water.

The precise nature of the products of the photolytic reaction is not known, but they may be envisaged as [O] and [H]. The former represents an oxidized material which in higher plants decomposes to yield molecular oxygen. The nature of this decomposition is still obscure, but it is known that it is accompanied by the formation of ATP. The reducing half of the product of photolysis interacts with NADP with the formation of $NADPH_2$ also with the simultaneous production of ATP. Thus the light-dependent reaction of photosynthesis provides the necessary requirements for the continuous operation of the so-called dark reactions, which result in the conversion of carbon dioxide into hexose. It is seen that although chlorophyll occupies the central role in trapping the energy of light, very many other components are necessary for the synthesis of carbon to carbon bonds. Without this elaborate machinery, non-photosynthetic organisms, including ourselves, would quickly use up all existing organic food. Without energy trapped from the sun, our own elaborate machinery for combusting carbon compounds would be useless.

The Chemistry of Living Processes 6

In the foregoing chapters an attempt has been made to summarize some facets of the chemical reactions of the cell and to show how they are woven into a highly organized system. It remains to be seen how the chemistry of the cell fulfils the function of sustaining the various processes of life. Space allows only a few examples to be discussed but these illustrate how attempts are being made to relate chemical events to biological processes.

6.1 Muscular contraction

One of the most obvious manifestations of the use of energy is its conversion into mechanical work. Besides the enzymes associated with glycolysis, muscle tissue contains two major protein constituents called myosin and actin. Under certain conditions of extraction, these two proteins are obtained complexed together in the form of actomyosin. Myosin and actin are the proteins which go to form the muscle fibres, or myofibrils and are closely associated with the contractile process. Both myosin and actin exhibit enzyme activity in that they catalyse the decomposition of ATP to give ADP and inorganic phosphate, and this suggests that the energy required for contraction of the myofibril is derived from ATP. Electron microscopic studies of myofibrils show that actin and myosin are closely associated in the intact muscle. Microscopic examination of voluntary muscle fibres reveals a cross-banded structure, which has been interpreted through electron microscopy as due to the interpenetration of alternating filaments of myosin and actin as shown diagrammatically in Fig. 6–1. The narrow, densest bands in the muscle correspond to the so-called Z-line seen in electron micrographs as the 'end plate' of discrete actin filaments. The broader bands of intermediate density represent segments of overlap between the actin and myosin filaments, the lightest zone, or H-zone corresponding to areas containing only myosin. In contracted muscle, this zone is much smaller and the process of contraction is believed to involve the sliding of the actin filaments so that a greater degree of overlap with myosin filaments results. That this process is associated with the decomposition of ATP seems fairly certain, since isolated muscle fibres contract if placed in a solution of ATP, but how the two processes are linked is largely conjectural at present. If actomyosin is added to a solution of ATP, not only is the ATP decomposed, but also the two proteins dissociate from each other and only re-associate when the hydrolysis of ATP is complete. The actomyosin complex most probably represents an artifact formed during extraction, and it is most unlikely that the organized association seen in intact fibrils survives the processes involved. However, the sliding together of the actin filaments brings a greater length of actin and myosin filaments in juxtaposition and it seems likely that the decomposition of ATP which accompanies muscular action facilitates the formation of linkages between the adjacent myosin and actin molecules.

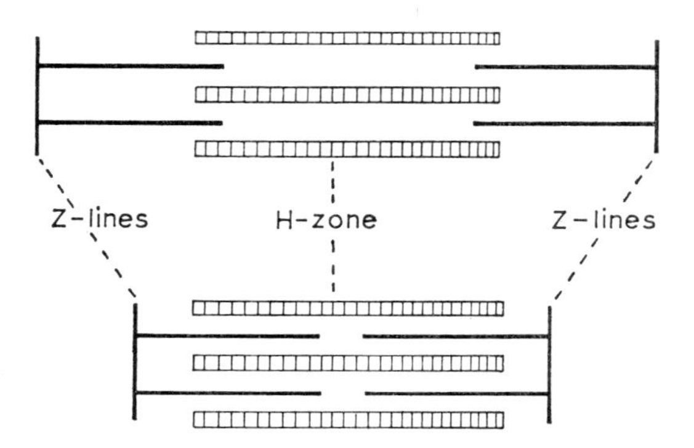

Fig. 6–1 Diagrammatic representation of extended (above) and contracted (below) muscle fibril. The heavy lines represent actin and the hatched parts myosin.

By taking muscle to pieces chemically and studying the components in detail, an outline is obtained showing how the chemical reactions in the muscle enable it to contract. The picture will not become clear, however, until means are found for recognizing the chemical reactions taking place *without* taking the muscle to pieces. Only then will it be possible to say precisely how actin and myosin interact *in vivo* and how ATP brings about this interaction.

6.2 Chemical reactions associated with vision

It has been known for a very long time that night blindness, the inability to see well in dim light, is relieved by eating raw liver. The efficiency of this treatment can now be ascribed to the increase in the supply of vitamin A which it provides. The human body is incapable of synthesizing vitamin A from small carbon compounds but is able to produce it by degradation of β-carotene, a pigment present in carrots. This conversion takes place in the intestine and the vitamin A so formed is stored largely in the liver. Carrots are just as effective as liver in preventing night-blindness.

The retina of the eye contains a pigment called visual purple or rhodopsin, which consists of a protein, opsin, complexed with the aldehyde

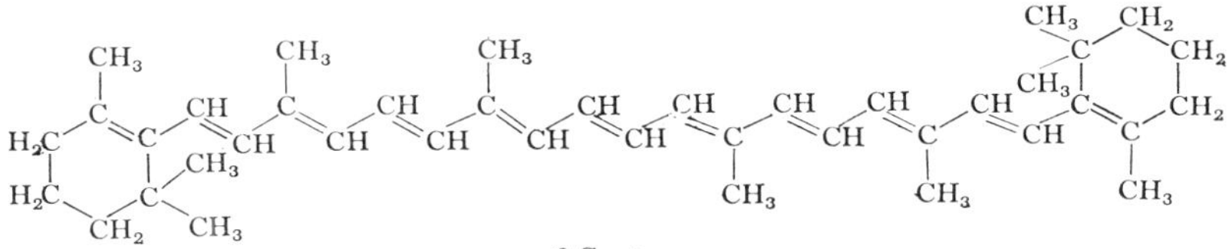

β-Carotene

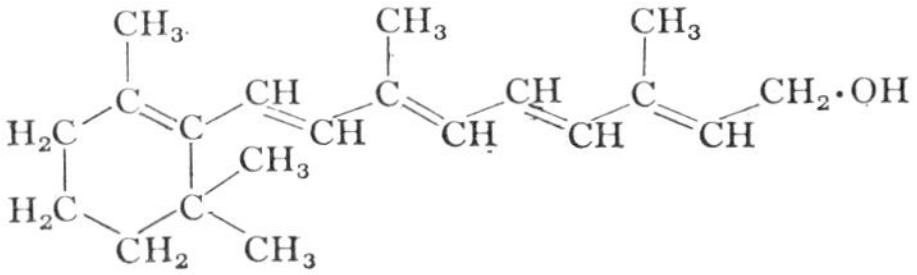

Vitamin A

corresponding to vitamin A, namely retinal. Opsin is colourless since it absorbs light only in the far ultra-violet region of the spectrum. Retinal is also almost colourless since it exhibits maximal absorption at a wave-length of 380 mμ, which is just beyond the visible range. On the other hand, rhodopsin is purple in colour, absorbing light around 500 mμ. When rhodopsin is irradiated with light, the colour is bleached, and the simultaneous increase in absorption at 380 mμ indicates that the fading of the colour is due to the decomposition of the rhodopsin into its components, namely opsin and retinal. Conversely, when retinal and opsin are incubated together in the dark, the formation of rhodopsin can be recognized by the increase in absorption at 500 mμ. The bleaching of rhodopsin can be seen to take place in the living retina, suggesting that it is concerned with vision.

In addition to the dissociation of the complex between opsin and retinal, a further change takes place in the latter molecule during the bleaching of rhodopsin. Compounds such as retinal, which contain double carbon to carbon bonds can exist in different forms known as geometrical, or *cis* and *trans* isomers. These arise because of the restriction of rotation around a double bond and result in different spatial distribution of atoms and groups attached to the doubly bound carbon atoms. Experiments with various isomers of retinal revealed that only one, namely, 11-*cis*-retinal (6-II) combines with opsin to give rhodopsin. The retinal produced by irradiation of rhodopsin was found to be 'all *trans*' retinal (6-I), showing that the retinal had undergone isomerization during the irradiation of the rhodopsin. It can be seen by comparing (6-I) and (6-II) that this isomerization involves a considerable change in the overall shape of the molecule and it is presumed that only the bent form of the molecule is able to fit together with the protein opsin. Clearly, before rhodopsin can be reformed, the isomerization of retinal must be reversed, the all-*trans* being converted into the 11-*cis* isomer. This isomerization is catalysed by an enzyme present in retina tissue known as retinal isomerase. Since *trans* isomers always have a lower energy content than *cis* isomers and are consequently more stable, the position of equilibrium in the isomerization favours the all-*trans* form. Because the 11-*cis* retinal is the only form able to combine with opsin, the equilibrium is disturbed, and the retinal is effectively trapped in the less stable form. The situation is analogous to the operation of a ratchet which enables the spring of a clock to be wound up into a shape having a higher energy content. The illumination of rhodopsin on the retina has an effect analogous to releasing the ratchet and allows the 11-*cis*-retinal to revert to the more stable form, with release of energy. What happens to this energy is unknown, but it seems almost certain that the bleaching of rhodopsin is one of the early chemical reactions resulting from the reception of light by the eye. It remains to be discovered whether the energy associated with 11-*cis*-retinal activates the optic nerve.

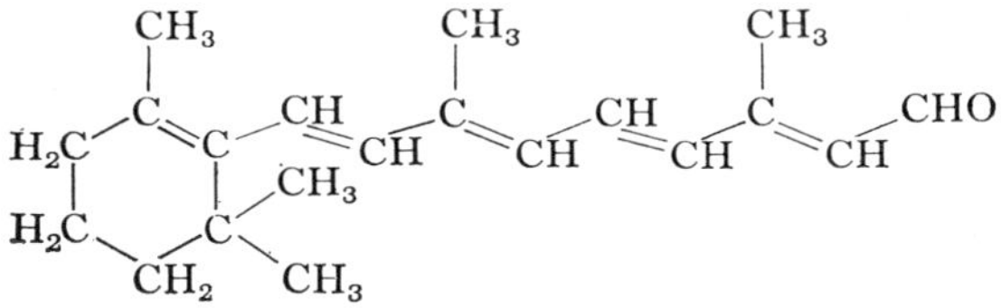

CH_3 CH_3 CH_3
C CH C CH C CHO
H_2C C CH CH CH CH
CH_3
H_2C C
CH_2 CH_3

6-I

CH_3 CH_3
C CH C CH
H_2C C CH CH CH
CH_3
H_2C C CH
CH_2 CH_3 CH_3 CH
CHO

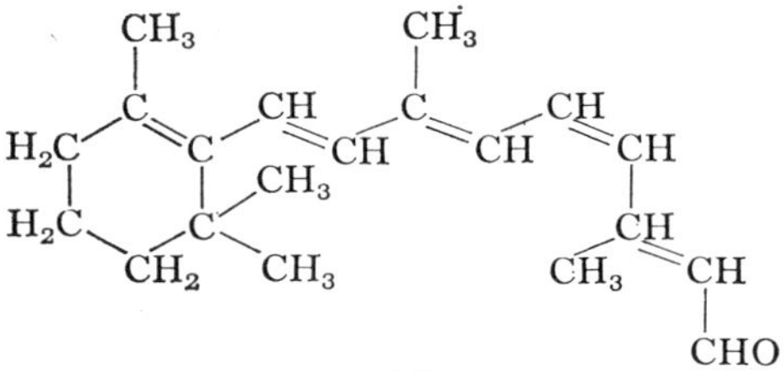

6-II

6.3 Some functions of the lipids

The term lipid is used to describe a large number of chemically diverse cell constituents having one feature in common, namely a large non-polar part to the molecule. Ionizable compounds are polar, and are able to associate readily with water molecules. In contrast, lipids either have no polar groups, or else so few that they are insignificant by comparison with the large hydrocarbon residues. There are three main kinds of lipid, the glyceride fats, the steroids and the carotenoids. β-Carotene typifies the carotenoids which have already been discussed in the previous section.

Tristearin is an example of a glyceride fat, being the ester of glycerol and the long-chain fatty acid stearic acid (6-III).

$$\begin{array}{l} CH_2\cdot O\cdot CO\cdot C_{17}H_{35} \\ \dot{C}H\cdot O\cdot CO\cdot C_{17}H_{35} \\ \dot{C}H_2\cdot O\cdot CO\cdot C_{17}H_{35} \end{array}$$

6-III

$$\begin{array}{l} CH_2\cdot O\cdot CO\cdot C_{17}H_{35} \\ \dot{C}H\cdot O\cdot CO\cdot C_{17}H_{35} \\ \dot{C}H_2\cdot O\cdot \underset{\dot{O}^-}{PO}\cdot O\cdot CH_2\cdot CH_2\cdot \overset{+}{N}(CH_3)_3 \end{array}$$

6-IV

No polar groups are present in the tristearin molecule, the major part of which resembles the aliphatic hydrocarbons. Accordingly, tristearin is insoluble in water. Some glyceride fats, known as phospholipids, differ in type from tristearin in that one of the hydroxyl residues of the glycerol is esterified with a nitrogen- and phosphorus-containing residue as shown in 6-IV. The presence of the ionized group in addition to the long hydrocarbon residues gives the compound a dual character.

Cholesterol (3-I, p. 26) and sodium cholate (6-V) are examples of steroids in which the non-polar structures are of a different type.

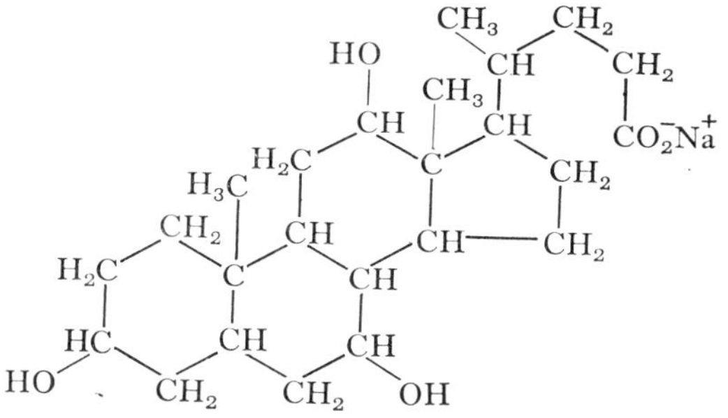

6-V

In the former, the effect of the single polar hydroxyl group is swamped by that of the large hydrocarbon residue, the overall properties being those of a non-polar molecule. Sodium cholate contains a completely ionized group and, like the phospholipids, exhibits both polar and non-polar characteristics.

The lipids play a number of diverse rôles in the living cell. The glyceride fats constitute a major source of energy because they are hydrolysed by enzymes known as lipases to yield glycerol and long-chain fatty acids. Glycerol is converted into glyceraldehyde 3-phosphate which is an intermediate in the glycolytic sequence of reactions. Long-chain fatty acids are oxidized with the removal of two carbon atoms at a time as acetic acid. This reacts with oxaloacetic acid to give citric acid which is metabolized as described in Chapter 5.

In contrast to the glyceride fats, cholesterol, the most widespread steroid in animal tissues, is not extensively broken down in cells and is not an important cellular fuel. Small amounts are used in the formation of steroid hormones which exert specific control over the chemical reactions of certain tissues and will be discussed briefly in the next section. Most of the cholesterol in food is converted into sodium cholate which is secreted in the bile and plays an important part in the absorption and digestion of fat.

Unlike non-polar lipids, which are immiscible with water, sodium cholate is soluble. The solution is colloidal in nature and contains aggregates of molecules in which the non-polar parts are in the centre and the polar parts on the outside as illustrated in Fig. 6–2(a). In this way, the non-

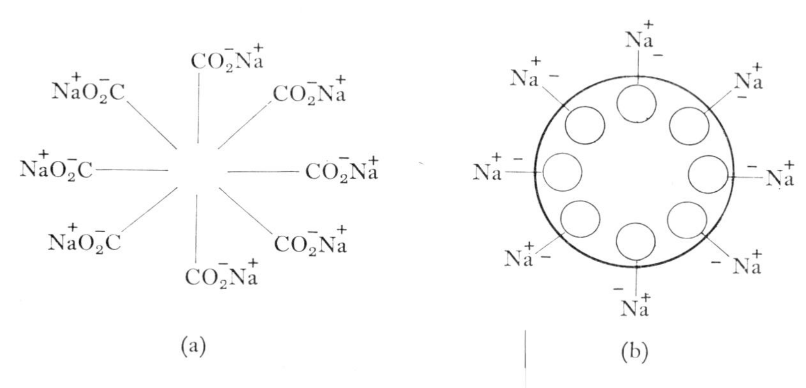

Fig. 6–2 (**a**) Aggregation of a polar lipid. (**b**) Stabilization of an oil droplet by a polar lipid.

polar residues are kept away from the water molecules which, on the other hand, are able to associate with the polar groups. A similar situation arises when a non-polar lipid is mixed with an aqueous solution of sodium cholate. In this case the sodium cholate molecules orientate themselves across the surface of the lipid droplet as shown in Fig. 6–2(b). This is the typical behaviour of a detergent and has the effect of stabilizing the droplet and so facilitating the formation of a colloidal solution.

One of the most important cellular functions of lipids is their participation in the structure of membranes. Not only are complex organisms divided into distinct organs, but the simplest cell contains structures which are bounded by membranes. Such membranes consist largely of proteins in close association with lipids. Both steroids and glycerides are present in membranes, but because of their dual characteristics, it is the phospholipoids which occupy the key position. Polar lipids are able to associate with the polar side-chains of proteins and form a water-resistant layer sandwiched between two layers of protein. It is believed that this waterproof lining to the protein part of the membrane is not continuous, but that it serves to channel the diffusion of water-soluble material to specific areas of the surface. Such areas act as penetrable holes, the shapes of which are determined by the structure of the protein layer. A number of cases are known in which membranes allow only certain molecules to diffuse into a cell and the proteins concerned in the process are classified as enzymes known as permeases. In their specificity, they resemble ordinary enzymes, but catalyse a process of diffusion instead of a chemical reaction. Permeases which have received most attention are those in the membranes of the bacterium *Escherichia coli* and the red blood cell. These enzymes control the penetration of lactose and glucose respectively. Disruption of a membrane makes it impossible for diffusion to take place in the normal way and this makes it difficult to characterize a permease in the form of a homogeneous lipoprotein. However, broken membranes have yielded protein fragments which specifically adsorb solutes to which the original whole membrane had been permeable and it is believed that these represent areas through which penetration occurred. Controlled diffusion plays a central role in many physiological events and its description in chemical terms promises to give a more detailed understanding of such processes in the future.

6.4 Growth and reproduction of the cell

Growth and reproduction are the foremost characteristics of life and it is of special interest to understand the chemistry of the processes involved. Growth of a bacterial culture consists of the repeated multiplication of cells, the progeny of which are identical with the parent cells. This entails the duplication of all cell constituents between each division. In higher organisms, growth is a more complex process in which differentiation as well as cell multiplication play a part.

Space does not permit detailed consideration of the synthesis of each cell constituent during growth. Methods by which they can be elucidated using isotopes have already been outlined. All the reactions involved in such syntheses are mediated by enzymes and for this reason, the key to the increase in cellular material associated with growth is the synthesis of proteins. The high degree of specificity exhibited by enzymes depends critically on the structure of the proteins concerned. It is therefore important to look particularly at the mechanisms for ensuring precise reproducibility of their structures.

If radioactively labelled amino acids are administered to animals and the tissues examined after a very short interval of time, the highest labelling is found in the ribosomes. Similar results are obtained with homogenates prepared from animal, plant and microbial tissue after incubation with labelled amino acids and it is concluded that protein synthesis takes place in the ribosome. However, fractionation of cell-free systems shows that the whole process requires a number of components located in various parts of the cell. The main ones are as follows:

(i) ATP;
(ii) Low molecular weight RNA present in the supernatant fraction, called soluble RNA or s-RNA;
(iii) Enzymes present in the supernatant fraction which are precipitated at pH 5 and referred to as pH 5 enzyme;
(iv) Ribosomes;
(v) High molecular weight RNA associated with ribosomes, known as messenger RNA or m-RNA.

By separating all these components from tissue homogenates it was possible to discover the sequence in which they participate in converting amino acids into protein. The first step is the reaction of a molecule of an amino acid with ATP with the elimination of a molecule of pyrophosphate and the formation of an amino acyl-AMP:

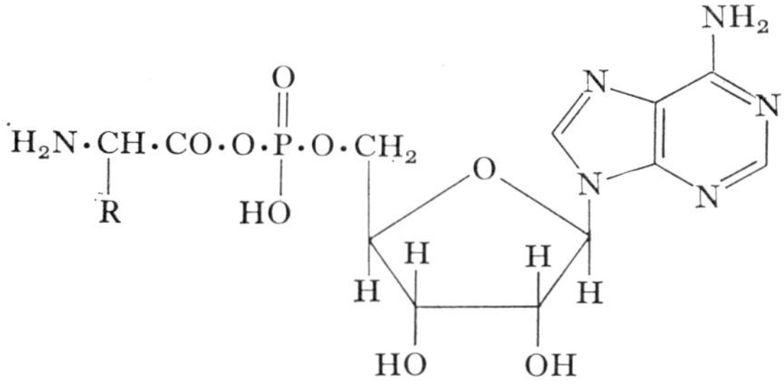

$$H_2N\cdot\underset{\underset{R}{|}}{CH}\cdot CO\cdot O\cdot\overset{\overset{O}{\|}}{\underset{\underset{HO}{|}}{P}}\cdot O\cdot CH_2$$

NH$_2$
N N
O
N N
H H
H H
HO OH

This type of compound is very reactive since it is a mixed acid anhydride between a carboxylic and a phosphoric acid derivative. In this sense, it is analogous to an acid chloride and as such is an acylating agent. Incubation of radioactively labelled amino acids with the supernatant fraction of tissue homogenates demonstrates the second stage of the sequence, which is the acylation of the terminal nucleotide residue of s-RNA by the acyl-AMP, AMP being liberated in the process:

$$H_2N\cdot\underset{\underset{R}{\cdot}}{CH}\cdot CO\cdot AMP + \text{s-RNA} \rightarrow H_2N\cdot\underset{\underset{R}{\cdot}}{CH}\cdot CO\cdot\text{s-RNA} + AMP$$

The exact position at which the amino acyl residue is attached is not known, but the structure at the end of the polynucleotide chain may be represented as shown on page 57. Both the reaction of amino acids with AMP and of the amino acyl-AMP with s-RNA require the presence of pH 5 enzyme.

If ribosomes are incubated with a mixture containing amino acyl-s-RNA both the amino acid and the s-RNA become bound to the ribosome. If protein is isolated from the ribosomes after the incubation, it is found that labelled amino acids originally combined with the s-RNA are now incorporated into the protein molecules. These results explain the initial

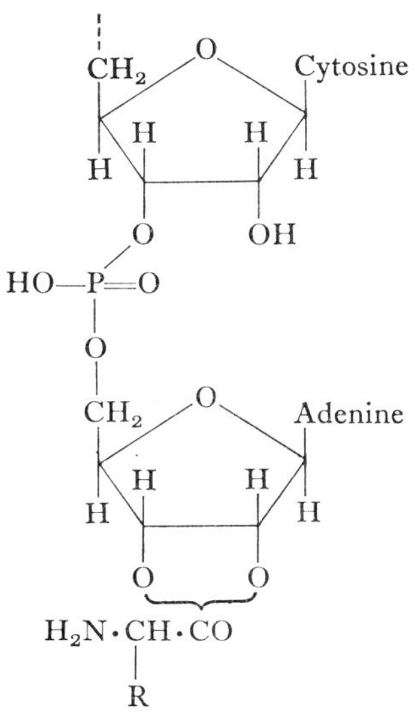

```
           ¦
           ¦
          CH2 ⁄ O ╲       Cytosine
           |                |
           H      H
          H                 H
                  |
           O     OH
          ⁄
   HO—P=O
           |
           O
           |
          CH2 ⁄ O ╲       Adenine
           |                |
           H      H
          H                 H
           |       |
           O       O
            ⎵⎵⎵⎵⎵
    H2N·CH·CO
         |
         R
```

appearance of labelled protein in the ribosomes in experiments with the complete cell-free system or with whole tissue. They also account for the requirement of component (i) to (iv) listed above. We now have to consider the part played by the fifth component.

Fractionation of s-RNA has revealed the presence of a number of different molecular species each of which combines with only one particular amino acid. Thus each amino acid is transferred to the site of protein synthesis on the ribosome carrying its specific s-RNA molecule by which it can be recognized before being added to the polypeptide chain. The overall structure of a protein is determined by the number and sequence of amino acid residues in its molecule and consequently the order in which amino acids are incorporated into a growing polypeptide chain is of supreme importance in the faithful duplication of protein molecules. The sequence in which amino acids carrying their s-RNA labels are accepted by the ribosomes is determined by messenger RNA. The existence of messenger RNA is an essential feature of modern concepts of chemical genetics which are based partly on experimental fact and partly on circumstantial evidence.

In studies on the epidemiology of pneumonia, Griffiths found that material present in dead cells of pathogenic pneumococci is able to transform live cultures of other strains which do not give rise to symptoms of the disease. Such transformed cells which have become pathogenic retain this character in subsequent generations. In 1944, AVERY, MCLEOD and MCCARTHY showed that DNA was responsible for the transformation to the genetically different strain. This demonstrated that the possession of a particular type of DNA by a cell confers on it certain inherited characteristics. Over a dozen other bacteria have been successfully transformed but no reliable evidence has been obtained of genetic transformation occurring in higher organisms. Thus transformation by DNA provides direct experimental evidence that the genetic message is associated with DNA in bacteria. A clear-cut demonstration that this holds good for other types of cell is not available, but the weight of circumstantial evidence is overwhelming.

The DNA of the cells of higher organisms is located mostly in the nucleus. On the other hand, protein synthesis takes place in the ribosomes which are located in the cytoplasm. If DNA controls the inherited reproducibility in protein synthesis, this separation of control and production poses a problem in communication. It was with this difficulty in mind that Jacob and Monod began a search for a messenger which could carry information from the control apparatus in the nucleus to the production unit in the cytoplasm. Once again, they turned to bacteria for the answer, although in bacteria a clearly distinct nucleus is not seen. For their search, they chose a so-called inducible bacterium in which the synthesis of a protein, the enzyme β-galactosidase, can be triggered off by addition of a substrate of the enzyme such as lactose. It can be as quickly stopped by removal of the substrate. They assumed that the information specifying the synthesis of such a protein would be rapidly formed and rapidly destroyed. Using isotopically labelled phosphorus, they found a very small fraction of cellular RNA which fulfilled these requirements in becoming labelled immediately after addition of lactose to a culture and being quickly destroyed after its removal. They concluded that a special part of the cellular RNA acts as messenger in transmitting information from control to production line. It remains to consider how genetic information is stored in DNA, how it is transferred by messenger RNA, and how it is translated into action in terms of the sequence of amino acids added to a growing polypeptide molecule.

In Chapter 2 it was shown how the doubly stranded structure of DNA provides a mechanism whereby replication of the molecule can be envisaged with the preservation of the sequence of nucleotide residues. From this it was a short step to equate faithful duplication of sequence with faithful retention of genetic information. Subsequently enzymes were found which catalyse the polymerization of ribonucleotides to RNA provided that DNA is present in the incubation mixture. The effect of the DNA is to determine the composition of the product RNA, in such a way that the percentage of guanine residues in the RNA equals that of cytosine residues in the DNA, and so on for the other pairs of complementary bases, uracil replacing thymine in the usual pairing. It was at once apparent that this enzymic reaction could well represent the synthesis of messenger RNA with DNA as template. The final link in the chain of events leading to the synthesis of protein according to a predetermined pattern concerns the delivery and interpretation of the message. Using radioactive labels, it has been demonstrated that RNA passes from the nucleus to the cytoplasm, which is interpreted as message in transit. Furthermore, electron microscopy shows a certain proportion of ribosomes which are grouped together like beads on a chain, with a thread linking them together. To interpret such threads as messenger RNA requires imagination rather than scientific knowledge, but in the realm of complex cell chemistry the reasonableness of a conclusion is often the chief criterion of its correctness. It is a matter for conjecture whether or not a thread-like structure in an electron micrograph is a molecule of RNA or some other macromolecule. But, fortunately, chemical experiments can be devised which show without doubt that protein synthesis in cell-free extracts is under the control of RNA. By adding artificially synthesized RNA molecules, the formation of polypeptide can be directed in a precise manner. Thus by addition of polyuridylic acid, that is an RNA in which all nucleotide residues carry uracil, ribosomes are made to produce polyphenylalanine and it is concluded that a sequence of uracil nucleotides in the RNA specifies a sequence of phenylalanine residues. As a result of many experiments with RNA with different nucleotide sequences, it has been concluded that each amino acid residue incorporated into a polypeptide chain is specified by a sequence of three nucleotide residues in the RNA added and therefore presumably by three nucleotides in a natural messenger RNA synthesized as a complementary copy of DNA. The formation of the messenger RNA is referred to as transcription of the information in the DNA molecule into the operational message written in a code of three-nucleotide characters. Some of the nucleotide triplets corresponding to amino acids are shown in Table 3. The final stage in protein synthesis is the translation of this message into a sequence of amino acids. The function of the label attached to each amino acid now becomes clear, since it is easy to envisage the pairing of, for example, a

Table 3 Some of the nucleotide sequences in m-RNA specifying amino acids incorporated into a growing polypeptide chain. A, G, U and C represent nucleotide residues derived from adenine, guanine, uracil and cytosine respectively.

Amino acid	*Nucleotide triplet*
Alanine	GCU
Cysteine	UGU
Glutamic acid	GAG
Glycine	GGG
Lysine	AAA
Phenylalanine	UUU
Serine	UCU
Tyrosine	UAU

U—U—U sequence in the messenger RNA with an A—A—A sequence in the s-RNA carried by phenylalanine. Thus a sequence in messenger RNA will result in the orderly arrangement of the corresponding amino acids carrying their labels.

The reconstruction of the mechanisms for the synthesis of protein according to a precise pattern hinges on a few crucial pieces of experimental evidence. Firstly the unequivocal demonstration that DNA carries genetic information in bacteria. Secondly that the synthesis of RNA in the cell is governed by DNA, and thirdly that the sequence of residues in a polypeptide is determined by a sequence of nucleotides in a molecule of

RNA. Unfortunately, as yet these processes cannot all be demonstrated experimentally in all cells. However, each new experiment carried out adds more weight to the hypothesis that this is the universal mechanism for the genetic control of cellular activities. All the genetic information is not transcribed in all cells and cell differentiation is believed to be the result of the activation of certain parts of the DNA in specialized cells. A chemical explanation of differentiation cannot yet be given, but experiments with some hormones suggest that their ability to stimulate specifically the synthesis of certain proteins is bound up with the formation of m-RNA. Many questions remain to be answered but it is certain that the interrelationships of DNA, RNA and protein synthesis hold the key to the understanding of many ramifications of the chemistry of the cell.

Further Reading

BRACHET, J. (1961). *The Living Cell.* Scientific American offprint No. 90. W. H. Freeman and Company, San Francisco.
This article provides a concise account of cell morphology which is increasingly important in interpreting the detailed chemistry of the cell.

CLARK, J. M. (Ed.) (1964). *Experimental Biochemistry*. Freeman, San Francisco.
Besides describing the detailed used of manometric techniques to which specific reference has been made, this book illustrates a number of experimental methods mentioned in the text.

FEINBERG, J. G. and SMITH, I. (1962). *Chromatography and Electrophoresis on Paper*. Shandon Scientific Company Ltd., London.
A number of experiments are described which can be carried out using simple apparatus.

STEINER, R. F. (1965). *The Chemical Foundations of Molecular Biology*. Van Nostrand, Princeton, New Jersey.
This is a more advanced text-book for biochemists but contains much which can be followed by serious students of biology.

TAYLOR, J. H. (Ed.) (1965). *Selected Papers on Molecular Genetics*. Academic Press, New York and London.
This publication presents in a compact form reprints of all the outstanding contributions to molecular genetics, together with annotations which make it possible to appreciate the significance of each landmark in relation to the historical development of the subject.

WATSON, J. D. (1965). *Molecular Biology of the Gene*. Benjamin, New York and Amsterdam.
A most readable account by one of the leading contributors to the subject.